The Theory of Fundamental Processes

ADVANCED BOOK CLASSICS
David Pines, Series Editor

Anderson, P.W., *Basic Notions of Condensed Matter Physics*
Bethe H. and Jackiw, R., *Intermediate Quantum Mechanics, Third Edition*
Feynman, R., *Photon-Hadron Interactions*
Feynman, R., *Quantum Electrodynamics*
Feynman, R., *Statistical Mechanics*
Feynman, R., *The Theory of Fundamental Processes*
Nozières, P., *Theory of Interacting Fermi Systems*
Pines, D., *The Many-Body Problem*
Quigg, C., *Gauge Theories of the Strong, Weak, and Electromagnetic Interactions*

The Theory of Fundamental Processes

Richard P. Feynman
late, California Institute of Technology

Advanced Book Program

CRC Press
Taylor & Francis Group
Boca Raton London New York

CRC Press is an imprint of the
Taylor & Francis Group, an **informa** business

First published 1961 by Westview Press

Published 2018 by CRC Press
Taylor & Francis Group
6000 Broken Sound Parkway NW, Suite 300
Boca Raton, FL 33487-2742

ISBN 13: 978-0-201-36077-6 (pbk)

Visit the Taylor & Francis Web site at
http://www.taylorandfrancis.com

and the CRC Press Web site at
http://www.crcpress.com

Cover design by Suzanne Heiser

Editor's Foreword

Addison-Wesley's *Frontiers in Physics* series has, since 1961, made it possible for leading physicists to communicate in coherent fashion their views of recent developments in the most exciting and active fields of physics—without having to devote the time and energy required to prepare a formal review or monograph. Indeed, throughout its nearly forty-year existence, the series has emphasized informality in both style and content, as well as pedagogical clarity. Over time, it was expected that these informal accounts would be replaced by more formal counterparts—textbooks or monographs—as the cutting-edge topics they treated gradually became integrated into the body of physics knowledge and reader interest dwindled. However, this has not proven to be the case for a number of the volumes in the series: Many works have remained in print on an on-demand basis, while others have such intrinsic value that the physics community has urged us to extend their life span.

The *Advanced Book Classics* series has been designed to meet this demand. It will keep in print those volumes in *Frontiers in Physics* or its sister series, *Lecture Notes and Supplements in Physics*, that continue to provide a unique account of a topic of lasting interest. And through a sizable printing, these classics will be made available at a comparatively modest cost to the reader.

These notes on Richard Feynman's lectures at Cornell on the Theory of Fundamental Processes were first published in 1961 as part of the first group of lecture note volumes to be included in the *Frontiers in Physics* series. As is the case with all of the Feynman lecture note volumes, the presentation in this work reflects his deep physical insight, the freshness and originality of his approach to understanding high energy physics, and the overall pedagogical wizardry of Richard Feynman. The notes provide both beginning students and experienced researchers with an invaluable introduction to fundamental processes in particle physics, and to Feynman's highly original approach to the topic.

David Pines
Urbana, Illinois
December 1997

Preface

These are notes on a special series of lectures given during a visit to Cornell University in 1958. When lecturing to a student body different from the one at your own institution there is an irresistible temptation to cut corners, omit difficult details, and experiment with teaching methods. Any wounds to the students' development caused by the peculiar point of view will be left behind as someone else's responsibility to heal.

That part of physics that we do understand today (electrodynamics, β decay, isotopic spin rules, strangeness) has a kind of simplicity which is often lost in the complex formulations believed to be necessary to ultimately understand the dynamics of strong interactions. To prepare oneself to be the theoretical physicist who will some day find the key to these strong interactions, it might be thought that a full knowledge of all these complicated formulations would be necessary. That may be so, but the exact opposite may also be so; it may be necessary to stay away from the corners where everyone else has already worked unsuccessfully. In any event, it is always a good idea to try to see how much or how little of our theoretical knowledge actually goes into the analysis of those situations which have been experimentally checked. This is necessary to get a clearer idea of what is essential in our present knowledge and what can be changed without serious conflict with experiments.

The theory of all those phenomena for which a more or less complete quantitative theory exists is described. There is one exception; the partial successes of dispersion theory in analyzing pion-nucleon scattering are omitted. This is mainly due to a lack of time; the course was given in 1959–1960 at Cal Tech, for which these notes were used as a partial reference. There, dispersion theory and the estimation of cross sections by dominant poles were additional topics for which, unfortunately, no notes were made.

These notes were made directly from the lectures at Cornell university by P. A. Carruthers and M. Nauenberg. Lectures 6 to 14 were originally written as a report for the Second Conference on Peaceful Uses of Atomic Energy, Geneva, 1958. They have been edited and corrected by H. T. Yura.

R. P. Feynman
Pasadena, California
November 1961

Contents

The Theory of Fundamental Processes

1 Review of the Principles of Quantum Mechanics

These lectures will cover all of physics. Since we believe that the behavior of systems of many particles can be understood in terms of the interactions of a small number of particles, we shall be concerned primarily with the latter. Bearing in mind that the present theories need modifications or revision to account for observed phenomena, we shall want to consider the foundation of quantum mechanics in their most general form. This is so we can get some idea of the minimum assumptions (and their character) which we use to formulate those parts of the theory we use in dealing with the new phenomena of the strange particles.

A rough outline of the book follows: First, we discuss the ideas of quantum mechanics, mainly the concept of amplitudes, emphasizing that other things such as the combination laws of angular momenta are largely consequences of this concept. Next, briefly, relativity and the idea of antiparticles. Following this, we give a complete qualitative description of all the known particles and all that is known about the couplings between them. After that, we return to a detailed quantitative study of the two couplings for which calculations can be carried out today; namely, the β-decay coupling and the electromagnetic coupling. The study of the latter is called quantum electrodynamics, and we shall spend most of our time with it.

Accordingly, we begin with a review of the principles of quantum mechanics. It has been found that all processes so far observed can be understood in terms of the following prescription: To every *process* there corresponds an *amplitude*†; with proper normalization the probability of the process is equal to the absolute square of this amplitude. The precise meaning of terms will become more clear from the examples that follow. Later we shall find rules for calculating amplitudes.

First, we consider in detail the double-slit experiment for electrons. A uniform beam of electrons of momentum p is incident on the double slit. To be more precise, we consider successive electrons, randomly distributed in the vertical direction (we prepare each electron with $p = p_x$, $p_y = p_z = 0$). (*Feynman:* They should come from a hole, at definite energy.)

†A complex number.

When the electron hits the screen we record the position of the hit. The process considered is *thus:* An electron with well-defined momentum somehow goes through the slit system and makes its way to the screen (Fig. 1-1). Now we are not allowed to ask which slit the electron went through unless

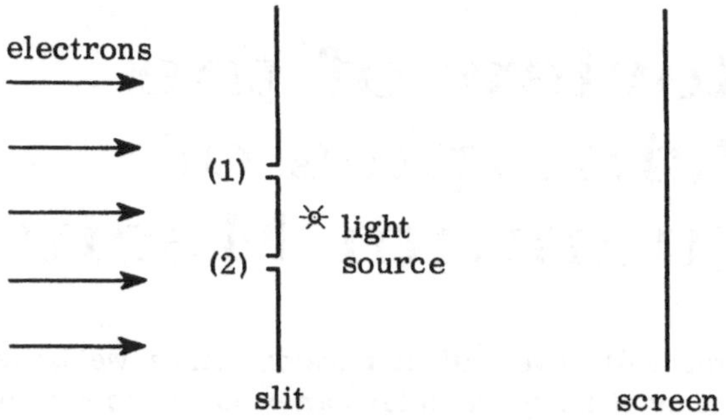

FIG. 1-1

we actually set up a device to determine whether or not it did. *But then we would be considering a different process!* However we can relate the amplitude of the considered process to the *separate amplitudes* for the electron to have gone through slit (1), (a_1), and through slit (2), (a_2). [For example, when slit (2) is closed the amplitude for the electron to hit the screen is a_1 (prob. $|a_1|^2$) etc.] Nature gives the following simple rule: $a = a_1 + a_2$. This is a special case of the principle of superposition in quantum mechanics (cf. reference 1). Thus the probability of an electron reaching the screen is $P_a = |a|^2 = |a_1 + a_2|^2$. Clearly, in general we have $P_a \neq P_{a_1} + P_{a_2}$($P_{a_1} = |a_1|^2$, $P_{a_2} = |a_2|^2$), as distinguished from the classical case. We speak of "interference" between the probabilities (see reference 2). The actual form of P_a is familiar from optics.

Now suppose we place a light source between slits 1 and 2 (see Fig. 1-1) to find out which slit the electron "really" did go through (we observe the scattered photon). In this case the interference pattern becomes identical to that of the two slits considered independently. One way of interpreting this situation is to say that the act of measurement, of the position of the electron imparts an uncertainty in the momentum (ΔP_y), at the same time changing the phase of the amplitude in an uncontrollable way, so that the average over many electrons yields zero for the "interference" terms, owing to the randomness of the uncontrollable phases (see Bohm[3] for details of this view). However, we prefer the following viewpoint: By looking at the electrons we have actually changed the process under consideration. Now we must consider the photon and its interaction with the electron. So we consider the following amplitudes:

a_{11} = amplitude that electron came through slit 1 and the photon was scattered behind slit 1

a_{21} = amplitude that electron came through slit 2 and the photon was scattered behind slit 1

a_{12} = amplitude that electron came through slit 1 and the photon was scattered behind slit 2

a_{22} = amplitude that electron came through slit 2 and the photon was scattered behind slit 2

The amplitude that an electron seen at slit 1 arrives at the screen is therefore $a' = a_{11} + a_{21}$; for an electron seen at slit 2, $a'' = a_{12} + a_{22}$. Evidently for a properly designed experiment $a_{12} \cong 0 \cong a_{21}$ so that $a_{11} \cong a_1$, $a_{22} \cong a_2$ of the previous experiment. Now the amplitudes a' and a'' correspond to different processes, so the probability of an electron arriving at the screen is $P'_a = |a'|^2 + |a''|^2 = |a_1|^2 + |a_2|^2$.

Another example is neutron scattering from crystals.

(1) Ignore spin: At the observation point the total amplitude equals the sum of the amplitudes for scattering from each atom. One gets the usual Bragg pattern.

(2) Spin effects: Suppose all atoms have spin up, the neutrons spin down (assume the atom spins are localized): (a) no spin flip—as before, (b) spin flip—no diffraction pattern shown even though the energy and wavelengths of the scattered waves are the same as in case a. The reason for this is simply that the atom which did the scattering has its spin flipped down; in principle we can distinguish it from the other atoms. In this case the scattering from atom i is a *different process* from the scattering by atom $j \neq i$.

If instead of (localized) spin flip of the atom we excite (unlocalized) spin waves with wavenumber $k = k_{inc} - k_{scatt}$, we can again expect some partial diffraction effects.

Consider scattering at 90° in the c.m. system [see Fig. 1-2 (a to d)]:

(a) Two identical spinless particles: There are two indistinguishable ways for scatter to occur. Here, total amplitude = 2a and $P = 4|a|^2$, which is twice what we expected classically.

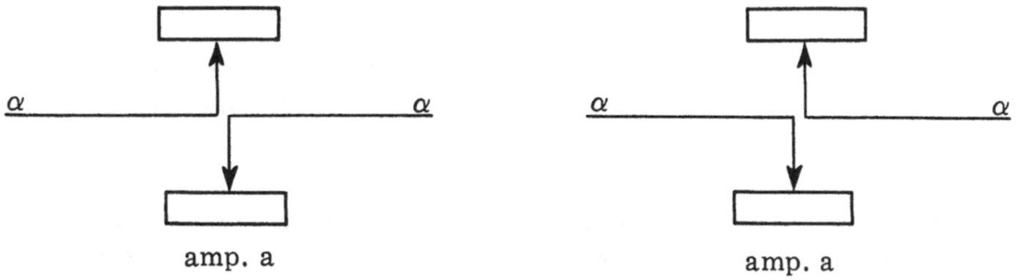

FIG. 1-2a

(b) Two distinguishable spinless particles. Here these processes are distinguishable, so that $P = |a|^2 + |a|^2 = 2|a|^2$.

(c) Two electrons with spin. Here these processes are distinguishable, so that $P = |a|^2 + |a|^2 = 2|a|^2$.

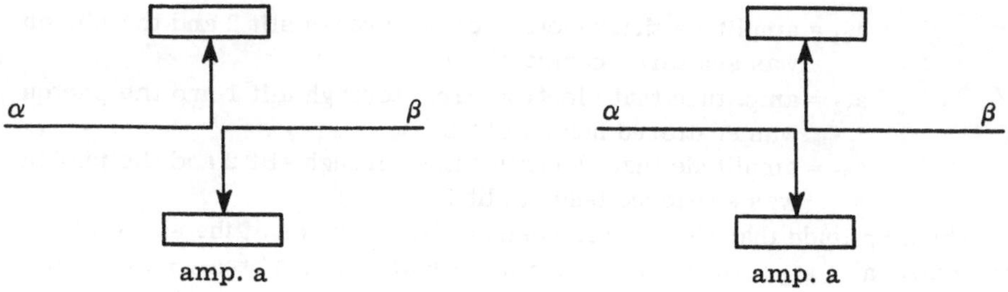

FIG. 1-2b

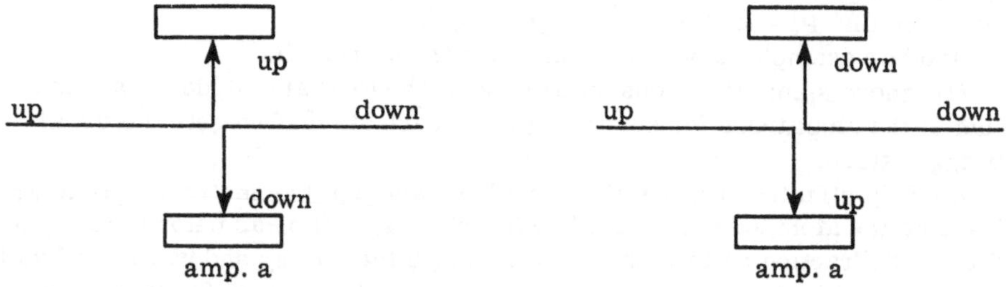

FIG. 1-2c

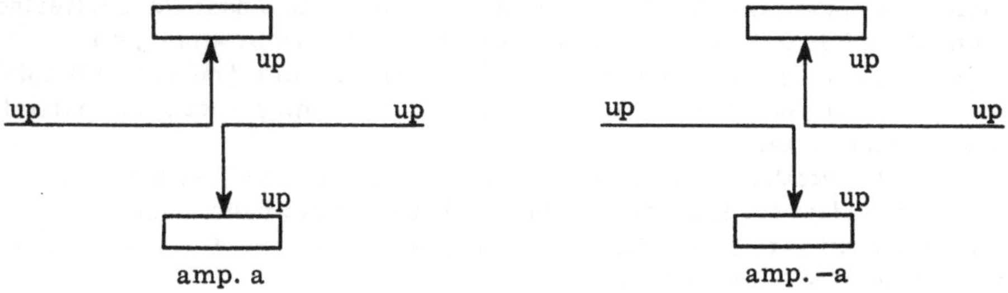

FIG. 1-2d

(d) But if both the incident electrons have spin up, the processes are indistinguishable. The total amplitude = a − a = 0. So here we have a new feature. We discuss this further in the next lecture.

Problem 1-1: Suppose we have two sources of radio waves (e.g., radio stars) and need to know how far apart they are. We measure this intensity in two receivers at the same time and record the product of the intensities as a function of their relative position. This measurement of the correlation permits the required distance to be computed. With one receiver there is no pattern on the average, because the relative phase of A and B sources is random and fluctuating. For example, in Fig. 1-3 we have put the receivers at a separation corresponding to that of two maxima of the pattern if the relative. phase is 0 (Table 1-1). If L and R are at separation between a maximum and a minimum we have Table 1-2. Thus find the probability of reception of photon coincidence in the counters. Examine the effect of changing the separation between the receivers. Consider the process from the point of view of quantum mechanics.

FIG. 1-3

TABLE 1-1

Relative phases of sources	L (common)	R (max)	Product
0°	2	2	4
180°	0	0	0
90°	1	1	1
270°	1	1	1
			Av. = 1.5

TABLE 1-2

Relative phases of sources	L (common)	R (max)	Product
0°	2	0	0
180°	0	2	0
90°	1	1	1
270°	1	1	1
			Av. = 0.5

Discussion of Problem 1-1. There are four ways in which we can have photon coincidences:

(1) Both photons come from A: amp. a_1.

(2) Both photons come from B: amp. a_2.

(3) Receiver L receives photon from A, R from B: amp. a_3.

(4) Receiver L receives photon from B, R from A: amp. a_4.

Processes (1) and (2) are distinguishable from each other and from (3) and (4). However, (3) and (4) are indistinguishable. [For instance, we could, in principle, measure the energy content of the emitters to find which had emitted the photon in case (1) and (2).]

Thus, $P = |a_1|^2 + |a_2|^2 + |a_3 + a_4|^2$. The term $|a_3 + a_4|^2$ contains the interference effects. Note that if we were examining electrons instead of photons the latter term would be $|a_3 - a_4|^2$.

2 Spin and Statistics

We should learn to think directly in terms of quantum mechanics. The only thing mysterious is why we must add the amplitudes, and the rule that $P = |\text{total amp.}|^2$ for a specific process. We return to consider the rules for adding amplitudes when the two alternative processes involve exchange of the two particles.

Consider a process P (amp. a) and the exchange process P_{ex} (amp. a_{ex}) *(indistinguishable from it)*. We find the following remarkable rule in nature: For one class of particles (called bosons) the total amplitude is $a + a_{ex}$; for another class (fermions) the total amplitude is $a - a_{ex}$. It turns out that particles with spin 1/2, 3/2, ... are fermions, and particles with spin 0, 1, 2, ... are bosons. This is deducible from quantum mechanics plus relativity plus something else. This is discussed in the literature by Pauli[4] and, more recently, by Lüders and Zumino.[5]

It is important to notice that, for this scheme to work, we must know all the states of which the particle (or system) is capable. For example, if we did not know about polarization we would not understand the lack of interference for different polarizations. If we discovered a failure of any of our laws (e.g., for some new particle) we would look for some new degree of freedom to completely specify the state.

Degeneracy. Consider a beam of light polarized in a given direction. Suppose we put the axis of an analyzer (e.g., polaroid, nicol prism) successively in two perpendicular directions, x and y, to measure the number of photons of corresponding polarization in the beam (x and y are of course perpendicular to the direction of the beam). Call the amplitude for the arrival of a photon with polarization in the x direction a_x, in the y direction a_y. Now, if we rotate the analyzer 45°, what is the amplitude $a_{45°}$ for arrival of a photon in that direction? We find that $a_{45°} = (1/\sqrt{2})(a_x + a_y)$; for a general angle θ (from the x axis) we have $a(\theta) = \cos\theta\, a_x + \sin\theta\, a_y$. The point is that only two numbers (here a_x and a_y) are required to specify the amplitude for any polarization state. We shall find this result to be connected intimately with the fact that any other choice of axes is equally valid for the description of the photon.

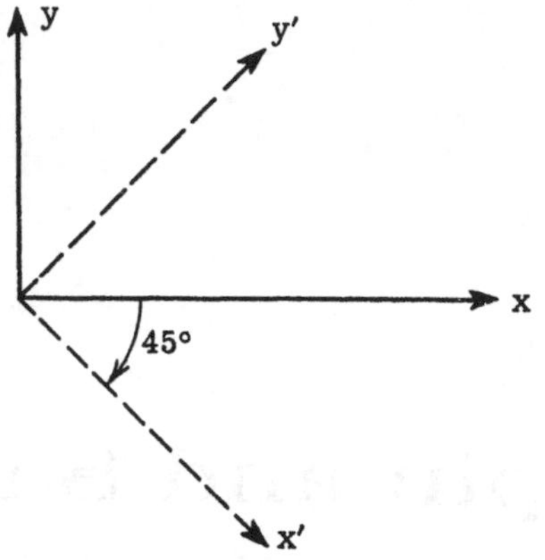

FIG. 2-1

For example (Fig. 2-1) consider the system of coordinates x′, y′ rotated −45° with respect to (x,y). An observer using this reference frame has

$$a'_{x'} = (a_x - a_y)/\sqrt{2}$$

$$a'_{y'} = (a_x + a_y)/\sqrt{2}$$

$$a'_{45°}(\text{in } x',y') = (a'_{x'} + a'_{y'})/\sqrt{2}$$

$$= [(a_x - a_y)/\sqrt{2}\,] + [(a_x + a_y)/\sqrt{2}\,]$$

$$= a_x \text{ (as it should be!)}$$

We could represent the state of the photon by a vector $\mathbf{e} = a_x\mathbf{i} + a_y\mathbf{j}$ in some two-dimensional space. Then the amplitude for the photon to be found with polarization in direction $\mathbf{v} = \mathbf{i}\cos\theta + \mathbf{j}\sin\theta$ is $\mathbf{e}\cdot\mathbf{v}$.

The hypothesis that the behavior of a system cannot depend on the orientation in space imposes great restrictions on the properties of the possible states. Consider (Fig. 2-2) a nucleus or an atom which emits a γ ray preferably along the z axis. Now rotate everything, nucleus plus detecting apparatus. We should expect that the photon is emitted in the corresponding direction.

If the nucleus could be characterized by a single amplitude, say, its energy, then the γ ray would have to be emitted with equal likelihood in all directions. Why? Because otherwise we could set things up so that the γ ray comes out in the x direction (for we can always rotate the apparatus, the working system; and the laws of physics do not depend on the direction of the axis). This is a different condition because the subsequent phenomenon (γ emission) is predicted differently. One amplitude for our state cannot yield two predictions. The system must be described by more ampli-

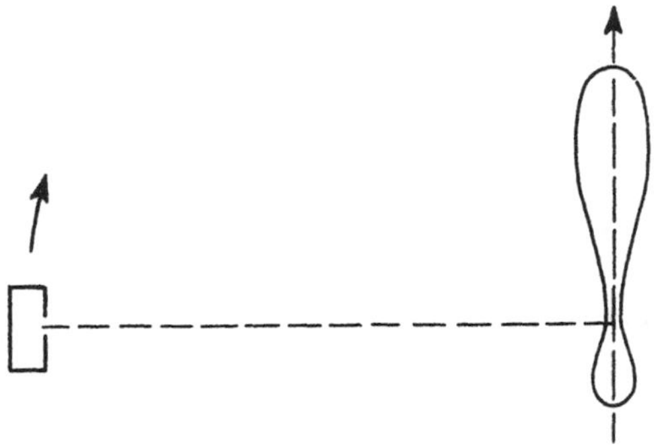

FIG. 2-2

tudes. If the angular distribution is very sharp we need a large number of amplitudes to characterize the state of the nucleus.

Suppose there are exactly n amplitudes which describe a system

$$\begin{pmatrix} a_1 \\ a_2 \\ \cdot \\ \cdot \\ a_n \end{pmatrix}$$

Now the problem: Suppose we know it is in the state $a_1 = 1,\ a_2 \cdots = a_n = 0$. After rotation what are the amplitudes characterizing the system in the new coordinates?

We define them as

$$\begin{pmatrix} D_{11}(R) \\ D_{21}(R) \\ \cdots\cdots \\ \cdots\cdots \\ \cdots\cdots \\ D_{n1}(R) \end{pmatrix}$$

Similarly if it starts in the state $a_2 = 1,\ a_1 = a_3 = \cdots = a_n = 0$, we have

$$\begin{pmatrix} D_{12}(R) \\ D_{22}(R) \\ \cdots\cdots \\ \cdots\cdots \\ \cdots\cdots \\ D_{n2}(R) \end{pmatrix}$$

Therefore we need an entire matrix $D_{ij}(R)$.

A more complicated case occurs if initially the system is in a state

$$\begin{pmatrix} a_1 \\ a_2 \\ \cdot \\ \cdot \\ \cdot \\ a_n \end{pmatrix}$$

After the rotation the new state is

$$\begin{pmatrix} a_1' \\ a_2' \\ \cdot \\ \cdot \\ \cdot \\ a_n' \end{pmatrix}$$

whereas $a_i' = \sum_j D_{ij}(R)a_j$. Think about why this is so.

3 Rotations and Angular Momentum

In the last lecture we spoke about an apparatus that produced an object in condition a:

$$a = \begin{pmatrix} a_1 \\ \cdot \\ \cdot \\ \cdot \\ a_n \end{pmatrix}$$

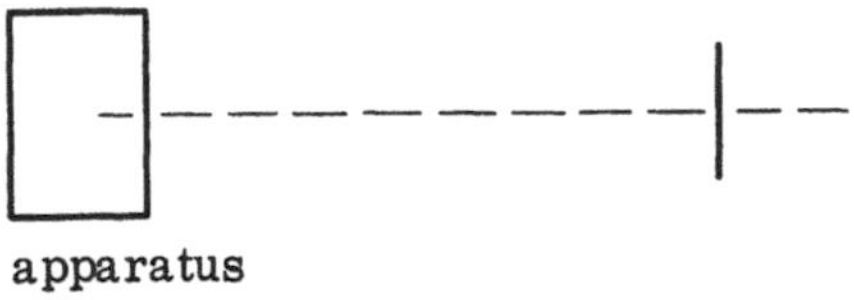

This requires further explanation; since we have introduced so far only the concept of an amplitude for the complete event: the production and detection of the object. This amplitude can be obtained as follows:

We assume that we have an amplitude b_i that the object produced is in some condition characterized by the index i. If it is in this condition, i, let a_i be the amplitude that it will activate some detector. Then the amplitude for the complete event (production and detection) is $a_i b_i$, summed over the possible intermediate conditions i.

Consider again the experiment of an electron passing through two slits

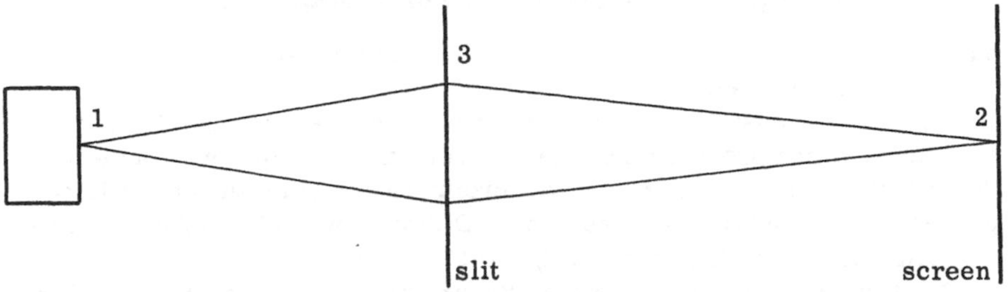

FIG. 3-1

(Fig. 3-1). If $a_{1\to 3}$ is the amplitude for an electron to go through one slit and $a_{3\to 2}$ the amplitude for an electron at this slit to reach the screen at 2, then the amplitude for the complete event is the product $a_{1\to 3} \times a_{3\to 2}$.

Now rotate the apparatus through $\mathbf{R}(|\mathbf{R}|$ = angle of rotation, $\mathbf{R}/|\mathbf{R}|$ = axis of rotation) so that the object is produced in condition a′ with respect to the fixed detector

$$a' = \begin{pmatrix} a_1 \\ a_2 \\ a_3 \\ \cdot \\ \cdot \\ a_n \end{pmatrix}$$

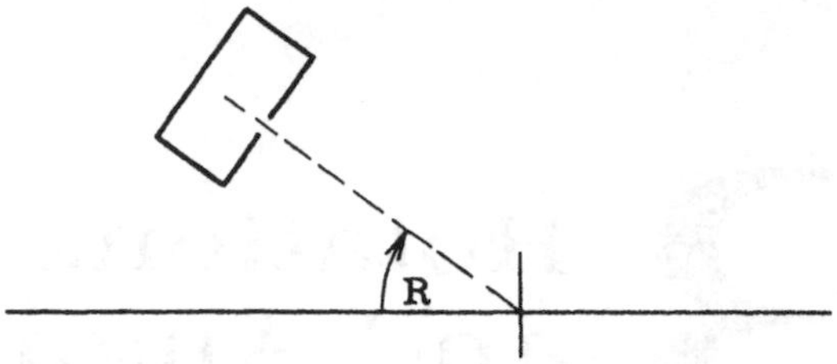

We have pointed out that this must be related to the a by an equation of the form $a' = D(R)a$, where the matrix $D(R)$ does not depend on the particular piece of apparatus. In another experiment (Fig. 3-2) we could have the same object produced in some other conditions b and b′. Then $b' = D(R)b$, and

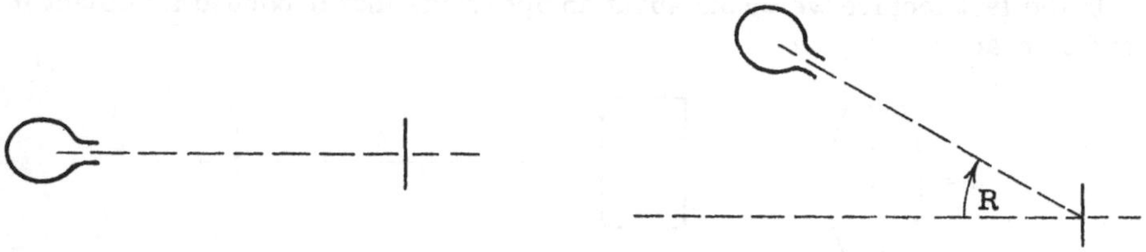

FIG. 3-2

the same D(R) is expected. Why must this relation be linear? Because objects can be made to interfere. Suppose we have two pieces of apparatus, one producing an object in condition a, the other producing the same object in condition b, and together producing it in condition $a + b$. After rotation we would have a′, b′, and also $a' + b'$, and also $a' + b'$, in order that the interference phenomena appear the same way in the rotated system. Then we have

$$a' = D(R)a \qquad b' = D(R)b \qquad (a + b)' = D(R)(a + b)$$

but $(a + b)' = a' + b'$, therefore $D(R)(a + b) = D(R)a + D(R)b$.

What else can we deduce?

Suppose we consider the apparatus that we rotated through R as a new apparatus, which produces the object in condition a′. Now we rotate it through S, as shown in Fig. 3-3. According to our rule, the object is now produced in a condition a″, where $a'' = D(S)a'$. Since $a' = D(R)a$, we have $a'' = D(S)D(R)a$, which means $D(SR) = D(S)D(R)$.†

Rotations form a group, and the D's are matrix representations of this group. It is by no means self-evident how to find them.

†Strictly speaking, we cannot prove that the amplitudes after rotation must be the same in both cases; only the squares must be the same. The amplitudes could differ by a phase factor. However, Wigner has shown that it could always be eliminated by redefining the D's.

Examples:

(1) An object represented by a single complex number. The D's are 1×1 matrices, i.e., a complex number can be chosen to be 1.

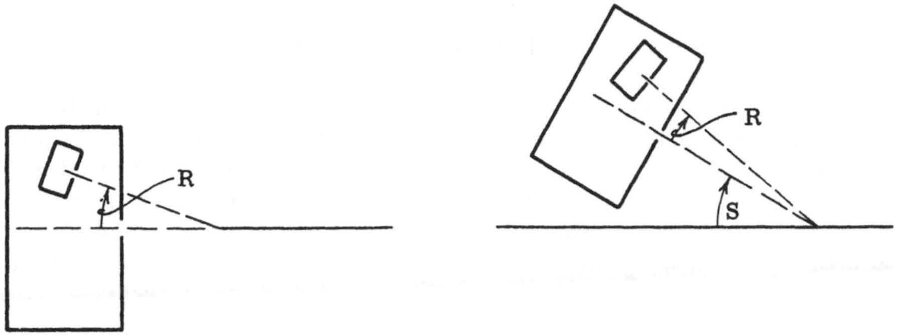

FIG. 3-3

(2) An object represented by a vector, hence by three amplitudes, the x,y,z components of the vector. The D's are the familiar matrices relating rotated coordinates.

Let us now go to the general analysis. Suppose we know a matrix for an infinitesimal rotation. Say, the rotation of 1° about the z axis. Then the rotation n° about the z axis is represented by

$$D(n^\circ \text{ around } z) = [D(1^\circ \text{ around } z)]^n$$

More generally, if we know $D(\varepsilon^\circ$ around z), then

$$D(\theta \text{ around } z) = [D(\varepsilon \text{ around } z)]^{\theta/\epsilon}$$

Now, if we rotate just a little we have approximately the identity, so to first order in ε, $D(\varepsilon \text{ around } z) = 1 + i\varepsilon M_z$. Also,

$$D(\varepsilon \text{ around } x) = 1 + i\varepsilon M_x$$

$$D(\varepsilon \text{ around } y) = 1 + i\varepsilon M_y$$

Now, we have $D(\theta \text{ around } z) = (1 + i\varepsilon M_z)^{\theta/\epsilon}$ and using the binomial expansion, one obtains, when $\varepsilon \to 0$,

$$D(\theta \text{ around } z) = 1 + i\theta M_z - \frac{\theta^2}{2!} M_z^2 - i\frac{\theta^3}{3!} M_z^3 + \cdots$$

which is often written $e^{i\theta M_z}$. The binomial expansion works, since M_z behaves like ordinary numbers under addition and multiplication.

If we want to rotate ε about an axis along the unit vector **v**, we find

$$D(\varepsilon \text{ around } \mathbf{v}) = 1 + i\varepsilon(v_x M_x + v_y M_y + v_z M_z)$$

and for a finite θ about $\mathbf{v}$,

$$D(\theta \text{ around } v) = \exp[i\theta(v_x M_x + v_y M_y + v_z M_z)]$$

But now we must be careful about the relative order of M_x, M_y, and M_z in the matrix products that appear in the series; these matrices do not commute. This follows from the fact that finite rotations do not commute. Consider the rotation of an eraser, Fig. 3-4 (a and b). (1) Rotate it 90° about the z axis and then 90° about the x axis (Fig. 3-4a); (2) rotate it 90° about the x axis, and then 90° about the z axis (Fig. 3-4b); and we get two entirely different results.

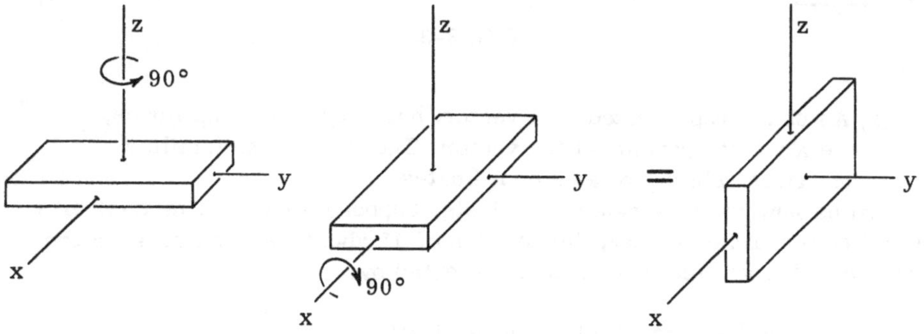

FIG. 3-4a

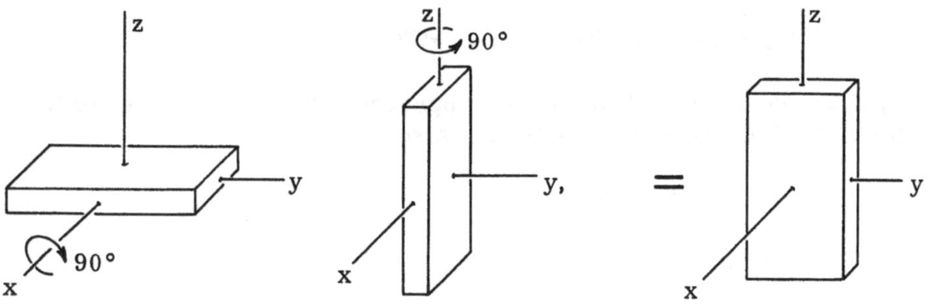

FIG. 3-4b

Let us discover the commutation relations between M_x and M_y. We consider a rotation ε about the x axis, followed by η about the y axis, then $-\varepsilon$ about the x axis and $-\eta$ about the y axis as in Fig. 3-5.

We follow the motion of a point starting on the y axis. Clearly the result is a second-order effect. It ends up just displaced by about $\varepsilon\eta$ toward the x

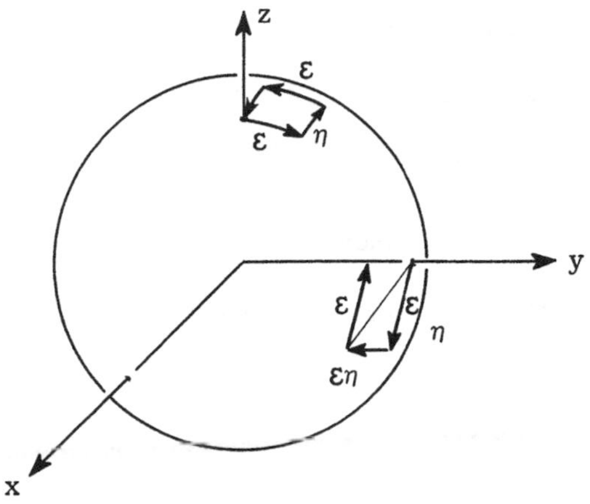

FIG. 3-5

axis. We note also that a point which starts on the z axis returns to the origin, and therefore the net displacement of the point on the sphere is just a rotation by an angle $\varepsilon\eta$ about the z axis. Keeping terms up to the second order, we have

$$[1 - i\eta M_y - \eta^2(1/2)M_y^2][1 - i\varepsilon M_x - \varepsilon^2(1/2)M_x^2][1 + i\eta M_y - \eta^2(1/2)M_y^2]$$
$$\times [1 + i\varepsilon M_x - \varepsilon^2(1/2)M_x^2] = 1 + i\varepsilon\eta M_z$$

Collecting coefficients of $\varepsilon\eta$ we find

$$M_xM_y - M_yX_x = iM_z$$

Similarly,

$$M_yM_z - M_zM_y = iM_x$$

$$M_zM_x - M_xM_z = iM_y$$

These are the rules of commutation for the matrices M_x, M_y, and M_z. Everything else can be derived from these rules. How this is done is given in detail in many books (e.g., Schiff). We give only a bare outline here. First we prove that $M_x^2 + M_y^2 + M_z^2 = M^2$ commutes with all M's. Then we can choose our a's so that they satisfy $M^2a = ka$, where k is some number. Construct

$$M_- = M_x - iM_y$$

and note

$$M_z M_- = M_-(M_z - 1)$$

Now, suppose $a^{(m)}$ satisfies

$$M_z a^{(m)} = m a^{(m)}$$

where m is another number; then

$$M_z b = M_z M_- a^{(m)} = M_-(M_z - 1) a^{(m)}$$

$$= (m - 1) M_- a^{(m)}$$

$$= (m - 1) b$$

Therefore,

$$b = c a^{(m-1)}$$

We normalize $a^{(m)}$ to unity; i.e.,

$$\sum_{j=1}^{n} a_j^{(m)*} a_j^{(m)} = 1 \qquad \text{for all } m$$

Therefore,

$$1 = (1/c^*c) \sum_n (M_- a^{(m)})_n^* (M_- a^{(m)})_n$$

$$= (1/c^*c) \sum_n a_n^{(m)*} (M_+ M_-) a^{(m)}$$

where $M_+ = M_x + iM_y$. Now

$$M_+ M_- = M_x^2 + M_y^2 + M_z$$

$$= M^2 - M_z^2 + M_z$$

and

$$M^2 a^{(m)} = k a^{(m)}$$

Therefore

$$c = [k - m(m - 1)]^{1/2}$$

Let $m = -j$ be the "last" state. How can we fail to get another if we operate by M_-? Only if $M_- a^{(-j)} = 0$ or $c = 0$ for $m = -j$, so $k = -j(-j - 1) = j(j + 1)$.

The same kind of steps (using M_+, which raises m by one, just like M_- lowers it) prove that if the largest value of m is $+j'$, then $k = j'(j' + 1)$, so that $j = j'$. Hence $2j'$ is an integer. The total number of states is $2j + 1$.

Examples:

(1) 1 state: $j = 0$

(2) 3 states: $j = 1$

m	Transforms like
1	$\frac{1}{\sqrt{2}}(x + iy)$
0	z
−1	$\frac{1}{\sqrt{2}}(x - iy)$

(3) 2 states: $j = 1/2$. This is a very interesting case. Let

$$a^{(1/2)} = \begin{pmatrix} 1 \\ 0 \end{pmatrix}$$

$$a^{(-1/2)} = \begin{pmatrix} 0 \\ 1 \end{pmatrix}$$

Using our general results we obtain

$$M_-\begin{pmatrix} 1 \\ 0 \end{pmatrix} = \begin{pmatrix} 0 \\ 1 \end{pmatrix}$$

since

$$[j(j+1) - m(m-1)]^{1/2} = [(1/2)(3/2) - (1/2)(1/2)]^{1/2} = 1$$

$$M_-\begin{pmatrix} 0 \\ 1 \end{pmatrix} = 0$$

Therefore,

$$M_- = \begin{pmatrix} 0 & 0 \\ 1 & 0 \end{pmatrix}$$

Likewise,

$$M_z\begin{pmatrix} 1 \\ 0 \end{pmatrix} = 1/2\begin{pmatrix} 1 \\ 0 \end{pmatrix}$$

$$M_z\begin{pmatrix} 0 \\ 1 \end{pmatrix} = -1/2\begin{pmatrix} 0 \\ 1 \end{pmatrix}$$

Therefore,

$$M_z = 1/2\begin{pmatrix} 1 & & 0 \\ & - & \\ 0 & & 1 \end{pmatrix}$$

Similarly we can show that

$$M_+ = \begin{pmatrix} 0 & 1 \\ 0 & 0 \end{pmatrix}$$

so that we can write

$$M_x = 1/2\begin{pmatrix} 0 & 1 \\ 1 & 0 \end{pmatrix} = (1/2)\sigma_x$$

$$M_y = 1/2\begin{pmatrix} 0 & -i \\ i & 0 \end{pmatrix} = (1/2)\sigma_y$$

$$M_z = 1/2\begin{pmatrix} 1 & 0 \\ 0 & -1 \end{pmatrix} = (1/2)\sigma_z$$

The above expressions also serve as the definition of the three important 2×2 matrices, the Pauli matrices σ_x, σ_y, σ_z. Check also that $\sigma_x^2 = \sigma_y^2 = \sigma_z^2 = 1$, $\sigma_x\sigma_y = -\sigma_y\sigma_x = i\sigma_z$. The main point of this is, that it all came out of nothing: that nature has no preferred axis and the nature of the principle of superposition were the only assumptions invoked.

However, we have made a very important hypothesis: We have assumed that the processes of production and detection are *well separated* and that in between one can talk of an amplitude that characterizes the object. This hypothesis has always been made (particularly in field theory) no matter how small the distance between the apparatus and the detector. It may turn out that it is not valid if these are too close together.

Another important assumption was to disregard any dynamic interference: There are no forces between our producing and measuring apparatus at least that are not describable by transfer of our object between them. An amplitude for two independent events is then also the product of the amplitude for each separate event.

Look at the example of the two stars A, B and the counters X, Y (Fig. 3-6). If $a_{B\to x}$ is the amplitude for the photon emitted at B to reach counter X and $a_{A\to y}$ is the corresponding amplitude for the photon emitted at A to reach counter Y, then $a = a_{B\to x} \times a_{A\to y}$ is the amplitude for occurrence of both events.

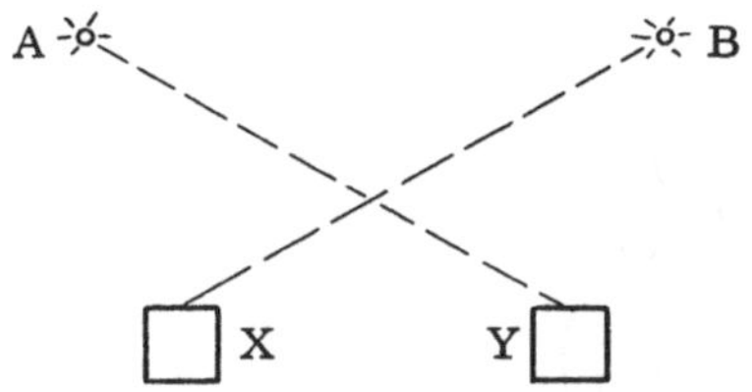

FIG. 3-6

4 Rules of Composition of Angular Momentum

A spin 1/2 state is characterized by two amplitudes. In general $a = a_+(1/2) + a_-(-1/2)$ where (1/2) stands for $\begin{pmatrix}1\\0\end{pmatrix}$, (−1/2) for $\begin{pmatrix}0\\1\end{pmatrix}$, and a for $\begin{pmatrix}a_+\\a_-\end{pmatrix}$.

For instance, the solution of

$$M_x a = (1/2)a$$

corresponding to spin up along the x axis is

$$a = (1/\sqrt{2})(1/2) + (1/\sqrt{2})(-1/2)$$

Also, down in x, $(1/\sqrt{2})(1/2) - (1/\sqrt{2})(-1/2)$; up in y, $(1/\sqrt{2})(1/2) + (i/\sqrt{2}) \times (-1/2)$; down in y, $(1/\sqrt{2})(1/2) - (i/\sqrt{2})(-1/2)$. In fact, it can be shown that every state represents spin in some direction.

Any system that has two complex numbers has an analogy in spin 1/2. For instance let us consider the polarization of light. Let x polarization be spin up and y polarization be spin down along an axis ζ in a "crazy" three-dimensional space. The other two axes we label ξ and η. Then spin up along ξ = 45° polarization; down, ξ = −45° polarization; up, η = RHC (right-hand circular polarization); down, η = LHC (left-hand circular polarization). If we draw a unit sphere centered at the origin of this space (Fig. 4-1), every state of polarization is represented by a point on it.

A general direction corresponds to elliptical polarization. Passing light through a 1/4-wave plate is a certain rotation. The connection between the polarization of light and direction in a three-dimensional space was exploited long ago by Stokes. It is very useful to understand certain processes, for example, masers. (The maser is a device using a system, the ammonia molecule, making transitions between two states under the influence of electric fields. Its analysis can be more easily understood by representing the state of the ammonia molecule at any time as a direction in some three-dimensional space, analogous to the ordinary space for a spin-1/2-electron.)

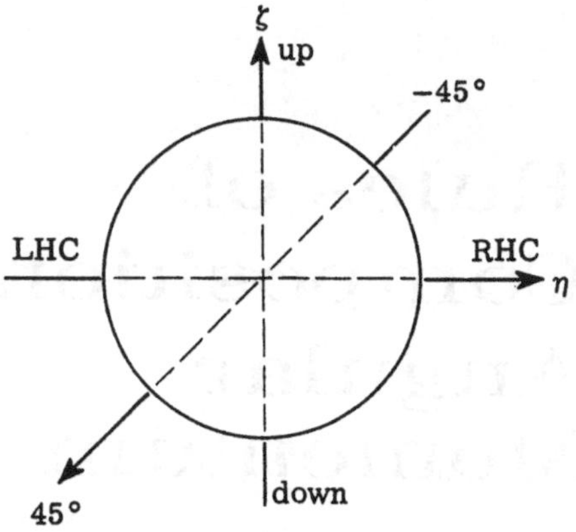

FIG. 4-1

Rules of Composition of Angular Momentum. Consider an apparatus that produces two particles which we label A and B. Suppose particle A has spin 1 and exists in three states with $m = +1, 0, -1$; and that particle B has spin 1/2 and exists in two states with $m = +1/2$ and $-1/2$. For each of A's three states, B can have two, so there are six possible states of the two particles together.

We may be thinking of an electron revolving around a nucleus. How do we characterize the combined system? We have matrices M_A and M_B which operate on the states ψ_A and ψ_B. Then

$$(1 + i\varepsilon M_z)\psi_A\psi_B = (1 + i\varepsilon M_{z_A})\psi_A(1 + i\varepsilon M_{z_B})\psi_B$$
$$= 1 + i\varepsilon(M_{z_A} + M_{z_B})\psi_A\psi_B$$

or†

$$M_z = M_{z_A} + M_{z_B}$$

The states of the combined system are given in Table 4-1. There are six states and one could jump to the conclusion that $j = 5/2$. However, there is no value of $m = \pm 5/2$ and also $m = \pm 1/2$ appears twice.

Actually $M^2 = (\mathbf{M}_A + \mathbf{M}_B)^2$ has two values for j:

$$j = 3/2 \qquad m = 3/2,\ 1/2,\ -1/2,\ -3/2$$

and

$$j = 1/2 \qquad m = 1/2,\ -1/2$$

†More precisely $M_z = M_{z_A} I_B + M_{z_B} I_A$ where I_A and I_B are the identity matrices for states A and B.

TABLE 4-1

m_A	m_B	m
1	1/2	3/2
0	1/2	1/2
1	−1/2	1/2
0	−1/2	−1/2
−1	1/2	−1/2
−1	−1/2	−3/2

Clearly the state $j = 3/2$, $m = 3/2$ is $(+1)(1/2)$. But which state corresponds to $j = 3/2$, $m = 1/2$? Recall

$$M_-(m) = [j(j+1) - m(m-1)]^{1/2}(m-1)$$

We have

$$M_- = M_-^A + M_-^B$$
$$M_-(1/2) = (-1/2)$$
$$M_-(-1/2) = 0$$
$$M_-(1) = \sqrt{2}\,(0)$$
$$M_-(0) = \sqrt{2}\,(-1)$$
$$M_-(-1) = 0$$

Then

$$M_-(1)(1/2) = \sqrt{2}\,(0)\,(1/2) + (1)(-1/2)$$

and

$$M_-(3/2, 3/2) = \sqrt{3}\,(3/2, 1/2)$$

Therefore,

$$(3/2, 1/2) = (\sqrt{2}/\sqrt{3})(0)\,(1/2) + (1/\sqrt{3})(1)\,(-1/2)$$

The state $(1/2, 1/2)$ is obtained by forming the linear combination of $(0)(1/2)$ and $(1)(-1/2)$, which is orthogonal to $(3/2, 1/2)$. We obtain the results given in Table 4-2.

TABLE 4-2

m	j = 3/2	j = 1/2
3/2	(1)(1/2)	
1/2	$(2/\sqrt{3})(0)(1/2) + (1/\sqrt{3})(1)(-1/2)$	$(1/\sqrt{3})(0)(1/2) - (2/\sqrt{3})(1)(-1/2)$
−1/2	$(2/\sqrt{3})(0)(-1/2) + (-1)(1/2)$	$-(1/\sqrt{3})(0)(-1/2) + (2/\sqrt{3})(-1)(1/2)$
−3/2	(−1)(−1/2)	

More examples: Add two spin = 1/2 states (Table 4-3) under exchange of spins. Now add two spin = 1 states (Table 4-4). For the addition of two equal angular momentum the biggest state is symmetric, the next antisymmetric, and so on.

TABLE 4-3

m	j = 1 (symmetrical)	j = 0 (antisymmetrical)
1	(1/2)(1/2)	
0	$(1/\sqrt{2})(1/2)(-1/2) + (1/\sqrt{2})(-1/2)(1/2)$	$\left\{(1/\sqrt{2})(1/2)(-1/2) - (1/\sqrt{2})(-1/2)(1/2)\right.$
−1	(−1/2)(−1/2)	

TABLE 4-4

m	j = 2 (symmetrical)	j = 1 (antisymmetrical)
2	(+1)(+1)	
1	$(1/\sqrt{2})[(+1)(0) + (0)(+1)]$	$(1/\sqrt{2})[(+1)(0) - (0)(+1)]$
0	$(1/\sqrt{6})[(+1)(-1) + (-1)(+1) + 2(0)(0)]$	$(1/\sqrt{2})[(+1)(-1) - (-1)(+1)$
−1	$(1/\sqrt{2})[(-1)(0) + (0)(-1)]$	$(1/\sqrt{2})[(0)(-1) - (-1)(0)]$
−2	(−1)(−1)	

m	j = 0 (symmetrical)
2	
1	
0	$(1/\sqrt{3})[(1)(-1) + (-1)(+1) - (0)(0)]$
−1	
−2	

Problem 4-1: Consider the addition of three spin = 1 angular momenta. Find the completely symmetric states. What total angular momenta occur?

5 Relativity

You are all familiar with the Lorentz transformation. For motion along the z axis the transformation equations between two Lorentz frames are

$$z' = (z - vt)/(1 - v^2)^{1/2} = z \cosh u - t \sinh u$$

$$t' = (t - vz)/(1 - v^2)^{1/2} = t \cosh u - z \sinh u$$

$$x' = x \qquad y' = y$$

where we have put $c = 1$ and introduced the quantity u (called "rapidity" by the experts!)

$$\tanh u = v/c$$

Note the equivalence of the second form of the transformation equations to a rotation through some imaginary angle. For transformations in the same direction rapidities are additive, i.e., if the rapidity between systems 1 and 2 is u, between 2 and 3 v, then the transformation from system 1 to 3 is characterized by the rapidity $w = u + v$. Transformations for different directions do not commute. The set of all Lorentz transformations (including rotations) form a group.

Problem 5-1: Suppose there exists an object with spin 1/2 in three dimensions; or consider a more-general state described by

$$a = \text{col } [a_1, a_2, \ldots, a_n]$$

What happens to a under Lorentz transformations?
Hint: What happens under rotation is the same as before. If you have time, also consider the problem of normalization.

Recall that the quantity $t^2 - x^2 - y^2 - z^2$ is invariant. We introduce the following notation: x_μ is the vector with components $x_4 = t$, $x_1 = x$, $x_2 = y$, $x_3 = z$.

If the vector $a_\mu = (a_4, a_1, a_2, a_3)$ transforms like x_μ then we call a_μ a four-vector. For example $p_\mu = (E,\mathbf{p})$ is a four-vector, with

$$E' = (E - vP_2)/(1 - v^2)^{1/2}, \text{ etc.}$$

The four-dimensional invariant scalar product of two four-vectors a_μ, b_μ, is

$$a \cdot b \equiv a_\mu b_\mu = a_y b_y - a_1 b_1 - a_2 b_2 - a_3 b_3$$

Introduce the quantity

$$\begin{aligned} \delta_{\mu\nu} &= 1 \qquad \mu = \nu = 4 \\ &= -1 \qquad \mu = \nu = 1,2,3 \\ &= 0 \qquad \mu \neq \nu \end{aligned}$$

Note that $\delta_{\mu\nu} a_\nu = a_\mu$. Also we define

$$\nabla_\mu \equiv (\partial/\partial t, -\nabla)$$

A useful invariant is $p \cdot p = E^2 - \mathbf{p} \cdot \mathbf{p} = m^2$. The skillful use of invariants in calculations often avoids making a Lorentz transformation. As a simple example, we consider p-p scattering: What is the minimum energy necessary to produce a proton-antiproton pair?

(E, p) ⟶ p (M, O) • p before (in the laboratory frame)

p. ·p • • p p after (in the center-of-mass frame)

$E = 4M \qquad p = 0$

We have $\mathcal{E}^2 - \mathbf{p}^2 = (4M)^2$. Hence $(E + M)^2 - \mathbf{p}^2 = 16\,M^2$, giving $E = 7M$. Thus the necessary kinetic energy is $6M \approx 5.6$ bev.

Waves. We know that a particle with energy-momentum p_μ has associated a wave $\varphi = u \exp(-ip_\mu x_\mu) = u \exp[-i(Et - \mathbf{p} \cdot \mathbf{x})]$. It is at once apparent that the phase of the wave is invariant under Lorentz transformation. This was, in fact, how DeBroglie found the relation between energy momentum and wavelength frequency. Problem 5-1 was to find how u transforms. Note that

$$\nabla_\mu = -ip_\mu \varphi$$

Positive and Negative Energies. The equation $E^2 = m^2 + \mathbf{p}^2$ has the two solutions

$$E = \pm (m^2 + \mathbf{p}^2)^{1/2}$$

It is a remarkable fact that we must take both solutions seriously. We find there are particles described by both the positive and negative frequency solutions. For $E > 0$, $\varphi \sim \exp(-iEt)$; for $E = -W$, $W > 0$, $\varphi \sim \exp(iWt)$. These two cases correspond to particles and antiparticles, respectively.

Represent the scattering of a classical particle by a space-time diagram, Fig. 5-1. (The shaded area represents the presence of an external potential that scatters the particle.)

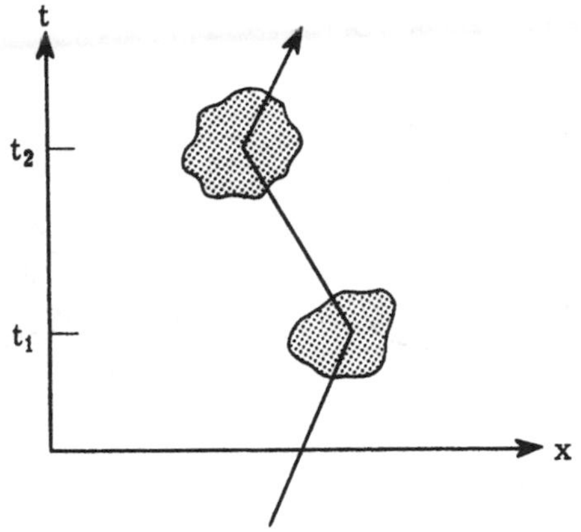

FIG. 5-1

Now suppose for a moment what would happen if the trajectories (or rather, in quantum mechanics, the waves) could go *backward in time!* Such a situation is shown in Fig. 5-2. Conventionally, this process appears as fol-

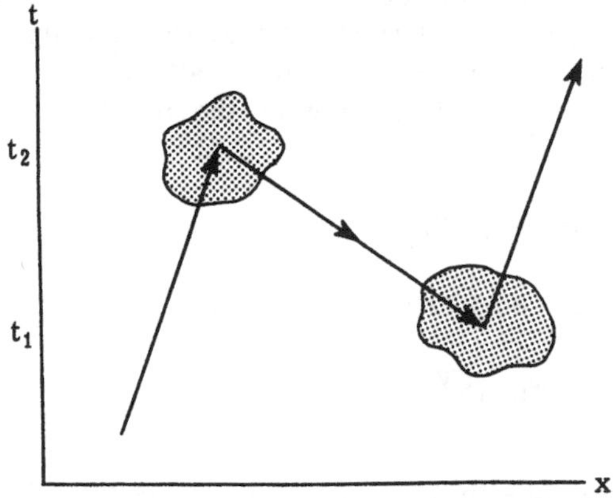

FIG. 5-2

lows (see Fig. 5-3): For $t < t_1$ there is only the incident electron. At t_1 an electron-positron pair is created by the external potential. At t_2 the positron annihilates the incident electron, so that for $t > t_2$ we have only the scattered electron. Instead we prefer to generalize the idea of scattering, so that the electron is considered to be scattered backward in time from t_2 to t_1. Then the conventional positron becomes an electron going backward in time. The two "double-scattering" processes then differ only in the time order of the successive scatterings, as we follow the electron along its world line. (One might think that this interpretation would imply that one could get information from the future; however, the full analysis shows that causality is not violated.)

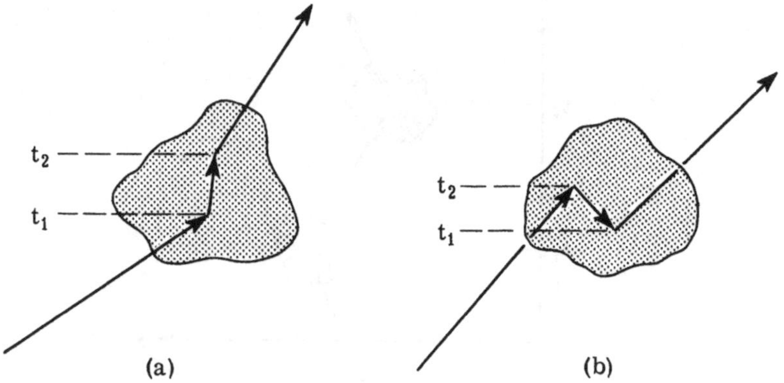

FIG. 5-3

Now we shall see how this viewpoint straightens out the difficulty of negative energies. We shall speak of initial (past) and final (future) states. We introduce the notion of *entry* and *exit* states (which have nothing to do with time). In the matrix element $\chi^* M\varphi$, φ is the *entry* state and χ the *exit* state. To determine *entry* and *exit* states we follow the world line of the particle even if we have to go backward in time. Considering the positron in Fig. 5-3, we see that for the exit state for scattering at t_2 we must use the initial state of the positron. For example, the matrix element for electron scattering is

$$\int \Psi^*_{\text{final}} M \Psi_{\text{initial}} \, dV$$

while that for positron scattering becomes

$$\int \varphi^*_{\text{initial}} M \varphi_{\text{final}} \, dV$$

The complete rule is thus:

Electrons: entry state in the matrix element is the initial state, and exit state in the matrix element is the final state;

Positrons: entry state in the matrix element is the final state, and exit state in the matrix element is the initial state.

As a testing example: Suppose an electron gives up some energy to the matrix machinery, $E_i = E_f + \omega > E_f$. Then the matrix element has the time dependence

$$\exp(iE_f t) \exp(i\omega t) \exp(-iE_i t)$$

[so we have shown that M varies like $\exp(i\omega t)$ if it extracts energy].

Now consider the positron case: $\exp(-iEt) = \exp(iWt)$. If the positron is treated in the old way (wrong!) the time dependence is

$$\exp(-iW_f t) \exp(i\omega t) \exp(iW_i t)$$

or $W_f = W_i + \omega$, i.e., the process creates energy!

But according to our correct prescription we should have

$$\exp(-iW_i t) \exp(i\omega t) \exp(iW_f t)$$

so that $W_i = W_f + \omega$ and the energy extracted by the M machinery shows as a loss to the positron, as we should like.

We can take more complicated cases: For example, the amplitude for annihilation of a pair is

$$(1) \quad \int \phi^*_{\text{initial positron}} M \varphi_{\text{initial electron}} \, dV$$

while the amplitude for pair creation is

$$(2) \quad \int \varphi^*_{\text{final electron}} M \varphi_{\text{final positron}} \, dV$$

It is easy to see that if M is energy absorbing, in (1) it soaks up all the energy while (2) gives zero, etc.

The first interpretation of the negative energy states was given by Dirac,

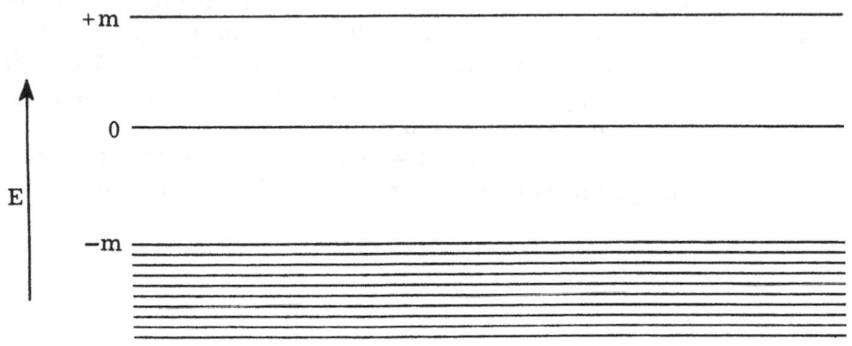

FIG. 5-4

who invoked the exclusion principle to prevent electrons from falling down to the negative energy states (Fig. 5-4). According to his picture, all negative energy states are filled up to $-mc^2$. This infinite sea is then unobservable. But we can see bubbles in the sea, i.e., the absence of an electron from a negative energy state. This bubble is then a positron.

For example, consider positron scattering: How could a positron get to a different state? An electron fills the initial hole, leaving a hole (positron) in the state† from which the electron jumped (Fig. 5-5). The matrix element for this process is

$$\int \varphi^*_{\text{initial}} \, M \, \varphi_{\text{final}} \, dV$$

which is the same result that the time-reversal argument gave.

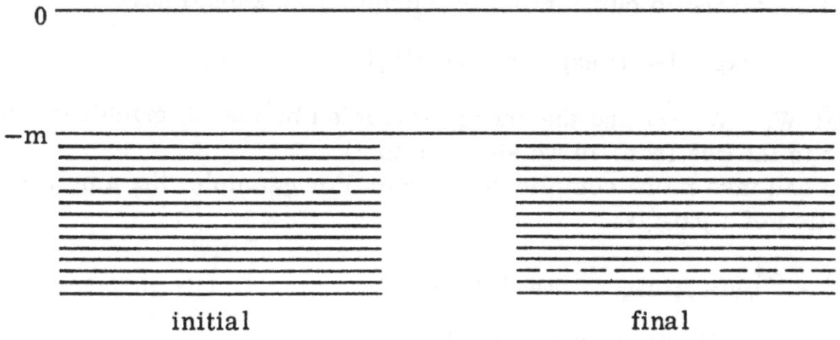

FIG. 5-5

There is a slight advantage in our formulation in that you do not have to deal with an infinite sea of electrons. But for bose particles, you could not fill the sea in a million years. It took eight years after the discovery of quantum mechanics before the Klein-Gordon equation was properly handled by Pauli and Weisskopf. Their prescription for interpreting the negative energy states of bosons was completely different from Dirac's theory and involves ideas of second quantization (Pauli and Weisskopf[6]). But their interpretation is again only equivalent to our rule that antiparticles simply reverse the role of entry and exit states.

(Lectures 6 to 14 are essentially the contents of an unpublished review article on strange particles by R. P. Feynman and M. Gell-Mann.)

†But, of course, having energy $-E = W$ and momentum $-p$.

6 Electromagnetic and Fermi Couplings

We want to describe how we stand today in the age-old attempt to explain all of nature in terms of the working of a few elements in infinite variety of combinations; in particular, what are the elements? The many unexpected results of high-energy experiments have revealed the incompleteness of our knowledge of these elements. We shall describe those theoretical ideas which have been most useful in partially sorting out this material. We shall concentrate on the ideas themselves and shall not have time to discuss carefully their origin or the history of their development. Furthermore, we can only describe how things seem at present. Every sentence might be prefaced by: Of course it might turn out to look quite different, but ... . We are well aware of the fragility and incompleteness of our present knowledge and of the manifold of speculative possibilities, but the exposition becomes awkward if we must continually refer to them. This is a survey of work from all over the world and not a report on any new contribution by the authors.

All the multitude of forms and the varieties of behavior of matter seem to be describable in terms of a limited number of fundamental particles interacting in definite ways. They obey the general principles of quantum mechanics and the principle of relativity. According to these principles, those of quantum field theory, there is nothing else besides particles. They have the intrinsic properties of rest mass and spin and a relation among one another described as coupling.

The Electromagnetic Coupling. Light, as an example, is represented by a particle, the photon of rest mass 0, spin 1. The emission of light by an excited atom is represented as the result of a fundamental coupling, or process $e \rightarrow e, \gamma$ (e for electron, γ for photon) meaning that there is a possibility (described more precisely by a mathematical quantity, an amplitude) that an electron may "become" a photon and an electron; the precise law of this coupling (how the amplitude depends on the directions of motion and spin of the particles concerned) is known very accurately (at least for energies less than 1 gev). When an electron in an atom does this, light is emitted by the atom. Each process implies its reverse at a corresponding amplitude; the arrows should be double-ended. The reverse occurs in light absorption.

$e \longleftrightarrow e,\gamma$ 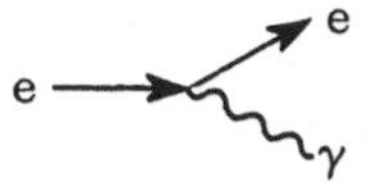 (6-1)

These relations are represented either by a double arrow, or by a diagram, the lines representing the particles coming in or out. The diagram is written to the right of Eq. (6-1).

A single free electron cannot emit one photon because of conservation of energy and momentum, but if two electrons are near one another, one may emit a photon which the other immediately absorbs. Quantum mechanics permits the temporary existence of states, which, if maintained, could not conserve energy. The penetration of a barrier in a decay of radioactive elements is a well-known example. The effect of this photon exchange we recognize in an interaction between the electrons, that is, as the electrical inverse-square repulsive forces. Thus all electric and magnetic forces among electrons, as well as the emission, scattering, and absorption of radio waves, light, and X-rays by electrons are described precisely and in detail by the simple law (6-1). The analysis of all this is called quantum electrodynamics.

A process only occurring by means of a temporary violation of energy conservation is called a virtual process. The diagram for electron interaction via virtual photon exchange is

$e,e \longleftrightarrow e,e$ indirect 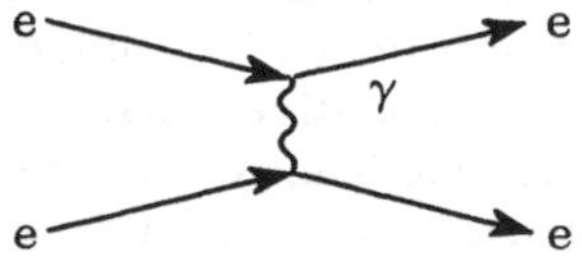 (6-2)

Only the real particles are represented by open-ended lines (two electrons in and out), the virtual photon has both its ends tied by the fundamental coupling (6-1).

Actually there is another principle, related to reversibility in time—that there are antiparticles to all particles. (For some neutral particles, like the photon, the antiparticle is the same as the particle.) The laws for this can be got by putting a particle on the other side of an equation and changing it to an antiparticle. The antielectron is the positron, so (6-1) implies

$e,\bar{e} \longleftrightarrow \gamma$ 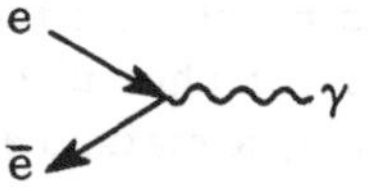

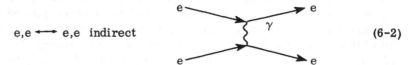

and in fact the laws of pair annihilation and creation are also completely specified by (6-1). Putting the photon on the other side of (6-1), $e,\bar{\gamma} \longleftrightarrow e$ only represents the same equation (6-1) again, for in fact the photon has no antiparticle, or more precisely the antiparticle to a photon is again a photon.

Other fundamental particles also couple with a photon; for example, if p represents a proton we have

$p \longleftrightarrow p, \gamma$ 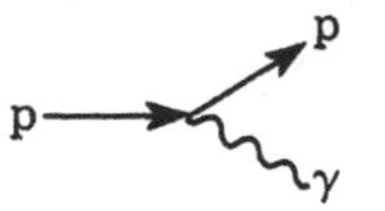

(6-3)

All particles that do so are called "charged." There are two remarkable laws about the numerical value of the charge e, neither of which is well understood. One is that all fundamental particles carry the same amount of charge (but it may be plus or minus). The other is that all other couplings are such that the total charge of all the particles in any reaction can never change. Finally the value of the charge e is measured in dimensionless form by the ratio $e^2/\hbar c = 1/137.039$. The value and origin of this number, which measures what we call the strength of the interaction (6-1) of electron with photon, are also mysterious. Its size has been determined empirically. Because 1/137 is a small number we say electrodynamics is a fairly weak interaction.

Fermi Couplings. Beside electrodynamics there is another even weaker coupling, Fermi coupling, required to account for the β decay of nuclei. For example, the decay of the neutron n into proton, electron, and antineutrino $n \rightarrow p, e, \bar{\nu}$ is taken to be the direct result of a coupling

$\bar{p}, n \longleftrightarrow \bar{\nu}, e$ 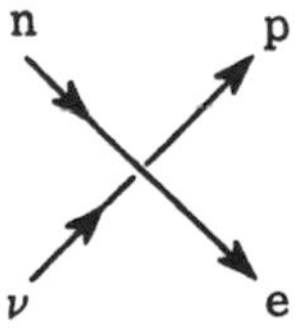

(6-4)

which we shall call a Fermi coupling. It is also often called the weak interaction. Another example of Fermi couplings is the decay of the muon μ, a charged particle just like an electron, but having mass 208.8 ± 1 times as much, $\mu \rightarrow e + \nu + \bar{\nu}$,

$\bar{\nu}, \mu \longleftrightarrow \bar{\nu}, e$ 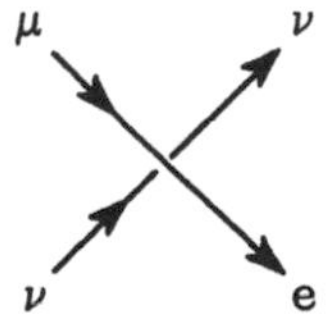

(6-5)

A third Fermi coupling is responsible for muon capture by nuclei $\mu + p \rightarrow n + \nu$,

$$\bar{\nu},\mu \longleftrightarrow \bar{p},n \qquad (6\text{-}6)$$

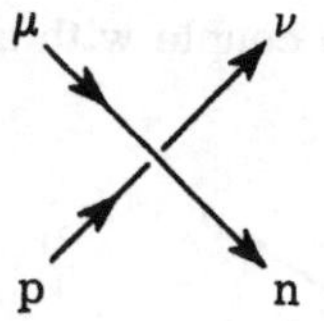

Others are undoubtedly involved in the slow decay of the strange particles that we shall discuss later.

7 Fermi Couplings and the Failure of Parity

Remarkably, the strength of the coupling for each of the three examples alone seems to be equal. It is measured by a constant G satisfying $GM^2/\hbar c = 1.01 \pm 0.01 \times 10^{-5}$, where M is the mass of the proton, included to make a dimensionless ratio. It is seen to be very small; the coupling very weak. The only couplings known for the neutrino (rest mass 0, spin 1/2) are Fermi couplings, so the interaction with matter of this particle is very small and its direct detection extremely difficult.

The detailed form of this Fermi coupling was established in 1957. It is remarkable in being the only one that violates the principle of reflection symmetry of physical laws (also known as the law of conservation of parity).

It was believed for a long time that for each physical process there existed (or could, in principle, exist) the mirror-image process. Consequently the distinction between right and left handedness was thought to be relative; neither could be defined in an absolute sense. Of course, we must leave out historical effects (for instance, the sense of rotation of our planet), because these correspond to a particular choice of initial conditions. If we radioed to an inhabitant of a different galaxy the most detailed instructions to construct a certain apparatus, he might end up building the mirror-image apparatus, since we could not communicate to him what our convention for right and left is. All experiments involving electromagnetic or nuclear forces completely supported this view.

This idea had in quantum mechanics the consequence of a property, called parity, of a state. Suppose that an apparatus produces an object in the state φ while the mirror-image apparatus produces the mirror object in state φ'. The principle of superposition requires that

$$\varphi' = P\varphi$$

where P is a linear operator. But

$$P\varphi' = \varphi$$

within the phase factor. Therefore $P^2 = 1$; and there are only two possible states:

$$\varphi \xrightarrow{\text{reflection}} +\varphi \quad \text{even parity}$$
$$\longrightarrow -\varphi \quad \text{odd parity}$$

The principle of reflection symmetry requires that a system remain in a state of a given parity for all times.

In early 1957, however, at the suggestion of Yang and Lee, a series of experiments in β decay were performed which violated this principle. Consider the Co^{60} experiment shown in Fig. 7-1. The Co^{60} spin was aligned at

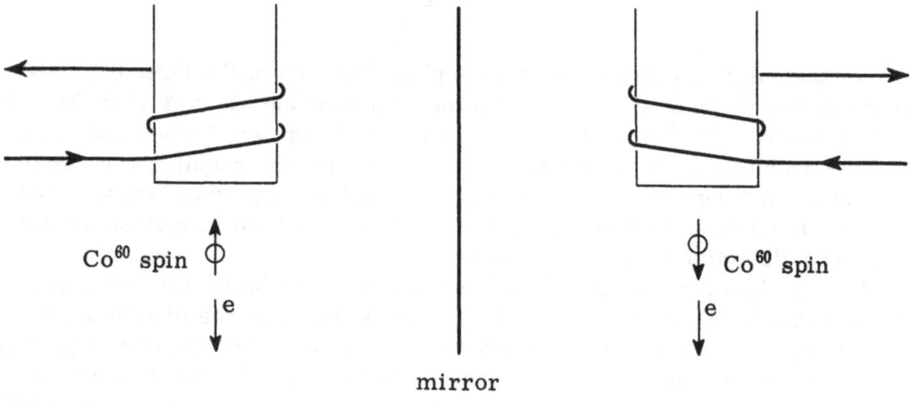

FIG. 7-1

very low temperatures in a magnetic field and the angular distribution of the emitted electrons observed. In turned out that the electrons came preferentially backward with respect to the Co^{60} spin. The mirror-image experiment indicates, however, that the electrons are emitted preferentially forward with respect to Co^{60} spin. It is therefore a physical process that cannot exist in our world. It seemed at last that we could radio our man in outer space how to distinguish right and left. We tell him to align some Co^{60} and define the direction of the magnetic field so that the electrons come out preferentially backward. But what if our fellow is made of antimatter, uses anti-Co^{60}, and observes positrons? At present our belief is that he will come up with the left-hand rule. That is, we think the mirror image is a possible world provided we also change matter into antimatter. Then, the positrons of anti-Co^{60} would be emitted parallel to the magnetic field.

Gravitation. In addition to these couplings there is one other, even-weaker, gravitation. The laws of gravitation, very satisfactorily known in the classical limit, have not yet been completely satisfactorily fitted into the ideas of quantum field theory, but presumably if it is done, there will be a particle (graviton, 0 rest mass, spin 2) coupled universally to every particle with a coupling constant so small that the gravitational force between electrons is 10^{-39} times the electrical force.

Nuclear Forces. Besides these weak couplings, gravitation, Fermi coupling, and electrodynamics, there must be some very much stronger. The forces binding neutrons together in the nucleus are much too strong to be explained otherwise. It is these strong couplings and the particles exhibiting them, which we wish now to discuss. There is no strong coupling to electron, muon, and neutrino (collectively called leptons or weakly interacting particles) nor to the photon or graviton. The particles that exhibit strong couplings are called "strongly interacting particles" and consist of the hyperons (among which are the neutron and proton) and the mesons. (The muon is technically not considered to be a meson.)

The force that holds the electrons around the nucleus is, of course, simply the electric attraction resulting from the virtual photon exchange between proton and electron implied by combining (6-1) and (6-3). But the nucleus is composed of neutrons and protons held together by a strong attractive force between them.

This nuclear force has been studied very carefully by studying nuclei and by scattering neutrons and protons by protons. It turns out to be not only much stronger than electrical force but also very much more complicated. In fact, except for one little unexpected thing, it turns out to be almost as complicated as it can be. Instead of the inverse-square force, it is a very strong repulsion at short distances, an attraction at somewhat larger distances falling rapidly to zero beyond 10^{-13} cm. The force depends on the relative spin directions of the p and n, and on the relation of these spin directions to the line joining the two particles. It even depends on the velocity of the particles and its relation to the spin direction (spin-orbit interaction). But, the one little unexpected thing: The force between p and p, that between p and n, and that between n and n all seem to be practically equal. Of course there are also electrical forces between p and p that do not have a counterpart in p and n, but when these are allowed for, by saying the total force is nuclear plus electrical, as nearly as we can tell the nuclear part of the p,p force, the p,n force, and the n,n force are all equal, when the particles are in corresponding states.

Isotopic Spin. Thus the origin of the strong nuclear forces has some kind of symmetry (called isotopic spin symmetry) of such a kind that it is irrelevant whether a particle is a proton or neutron. Even the small rest mass difference between neutron and proton most likely is the mass associated with the electromagnetic field surrounding these particles.

We learn our first lesson. Strongly interacting particles come in sets. The nucleons are a set of two, proton and neutron. We say the single nucleon has two states, the proton and the neutron. These states have the same energy. It is analogous to the two spin states of an electron spin "up" and "down" along some axis, which states have the same energy if no magnetic field is present.

In fact because the quantum mechanics of a two-state system is so thoroughly known to theoretical physicists from a study of a spin-1/2 system, they like to make full use of the analogy and to say the two states of the nucleon represent the "up" and "down" states of an object of spin 1/2 in an imaginary three-dimensional space. This space is called isotopic spin space. We say the nucleon has isotopic spin 1/2. The equality of the forces results from the hypothesis that the direction of the axis can be taken to be in any direction in isotopic spin space. In ordinary space this possibility of choosing the axis arbitrarily has a consequence the conservation of angular momentum. The strong couplings then satisfy a corresponding law, the law of conservation of total isotopic spin.

Since a particle of spin 0, 1/2, 1, 3/2, ... has 1,2,3,4, ... states, respectively, when we find that say three particles form a set, we say it has isotopic spin 1, etc. Then we can use the known laws of combining states of different angular momenta, to determine how these particles may be coupled to each other to preserve the symmetry between proton and neutron, or more generally, as we say, to preserve isotopic spin symmetry.

This applies only to the strong couplings; the isotopic spin conservation is destroyed by weak interactions like electrodynamics. A proton and a neutron have, of course, completely different couplings to a photon.

The Pion-Nucleon Coupling. To give an example of these ideas, suppose, as Yukawa suggested, that the nuclear forces are the result of a process analogous to (6-2) but instead of the electron there stands a nucleon and in place of the photon there is another particle. Let us try to do this. Suppose there is a fundamental process analogous to (6-1), such as

$p \longleftrightarrow n, \pi^+$

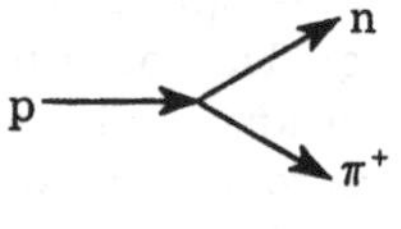

(7-1)

where π^+ stands for a positive pion, a new particle, carrying positive charge to keep the law of conservation of charge intact. Now there would be a force between a proton and neutron by the virtual transfer of a π^+:

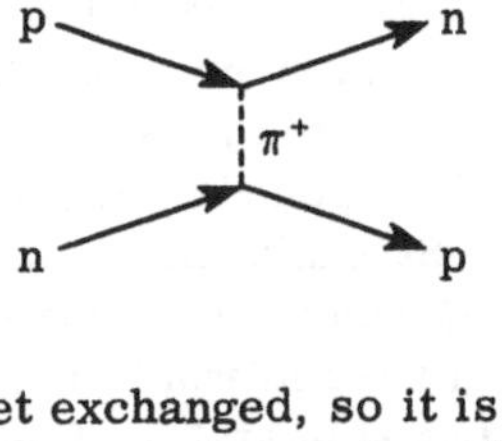

(Incidentally the p and n get exchanged, so it is called an exchange force.) Between two protons Eq. (7-1) can also make a force by exchanging two

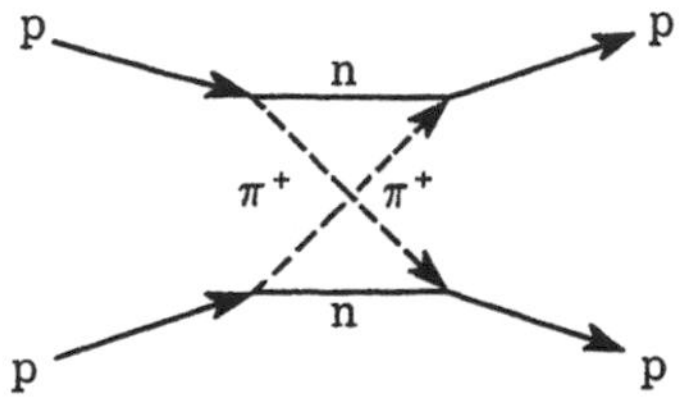

pions. But the force from the exchange of two cannot give the same answer as the force resulting from the exchange of one. So there must be some way for two protons to exchange one pion. There must be a neutral pion and a process like

$p \longleftrightarrow p, \pi^\circ$ 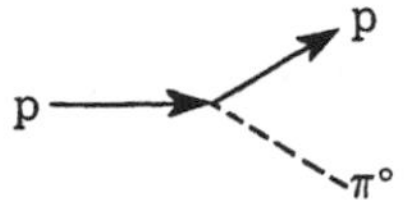 (7-2)

and also for neutrons,

$n \longleftrightarrow n, \pi^\circ$ 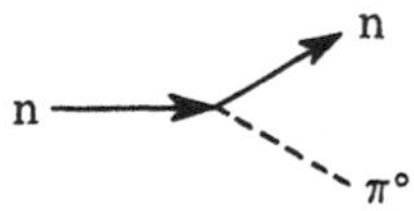 (7-3)

The new possibilities change the n,p force, too, giving a nonexchange part:

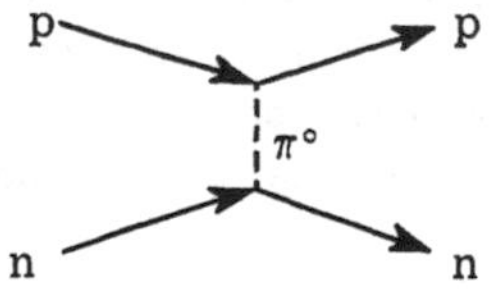

After some trials it is found that if (7-1) is true, (7-2), (7-3), and (7-4) are all necessary, but the amplitude (the analogue of electric charge, but for pion couplings) for (7-1) and (7-4) must both equal $\sqrt{2}$ times (7-2) and that of (7-3) equals (7-2) but of opposite sign. Then one can show that the symmetry of nuclear forces pp = pn = nn will remain true in all circumstances and no matter how many pions are exchanged:

$n \longleftrightarrow p, \pi^-$ 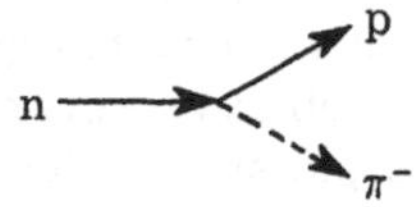 (7-4)

8 Pion-Nucleon Coupling

We assume the three fundamental interactions

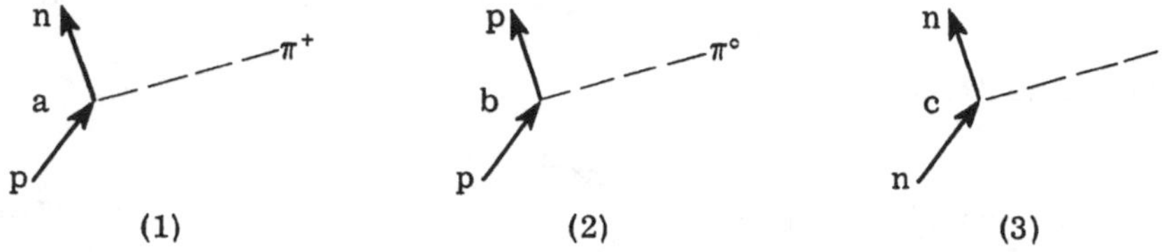

where a, b, and c are the amplitudes for processes (1), (2), and (3), respectively, and we want to determine the coupling constants a, b, and c which give rise to the symmetry in nuclear force (p,p) = (n,n) = (p,n) when they are in corresponding states. In lowest order we have the processes of Fig. 8-1. We must therefore have $bc + a^2 = b^2 = c^2$. If $b = c$ $a = 0$ there is no interaction between the charged π and nucleons contrary to experimental fact. Consequently,

$$b = -c \qquad a = (2b)^{1/2}$$

The choice $a = -(2b)^{1/2}$ corresponds to a different definition of the π-meson phase, which is arbitrary anyway.

But this result is also easily gained another way. If we have a triplet pion (π^+, π°, π^-) it is isotopic spin 1. In the reaction $N \leftrightarrow N + \pi$ (we represent a nucleon, neutron, or proton by N) the left side, a single nucleon, has isotopic spin 1/2. The right side has six states $(p\pi^+)(p\pi^\circ)(p\pi^-)(n\pi^+)(n\pi^\circ)(n\pi^-)$ but these may be analyzed into a doublet and quartet, for we combine an isotopic spin 1/2 nucleon and an isotopic spin 1 pion and can make up isotopic spin 1/2 and 3/2. If isotopic spin is conserved in strong couplings, the state on the right must be isotopic spin 1/2, too, and by analogy to laws coupling angular momentum we deduce

$$p \longleftrightarrow (p,\pi^\circ) + \sqrt{2}\,(n,\pi^+) \tag{8-1}$$

$$n \longleftrightarrow (n,\pi^\circ) + \sqrt{2}\,(p,\pi^-) \tag{8-2}$$

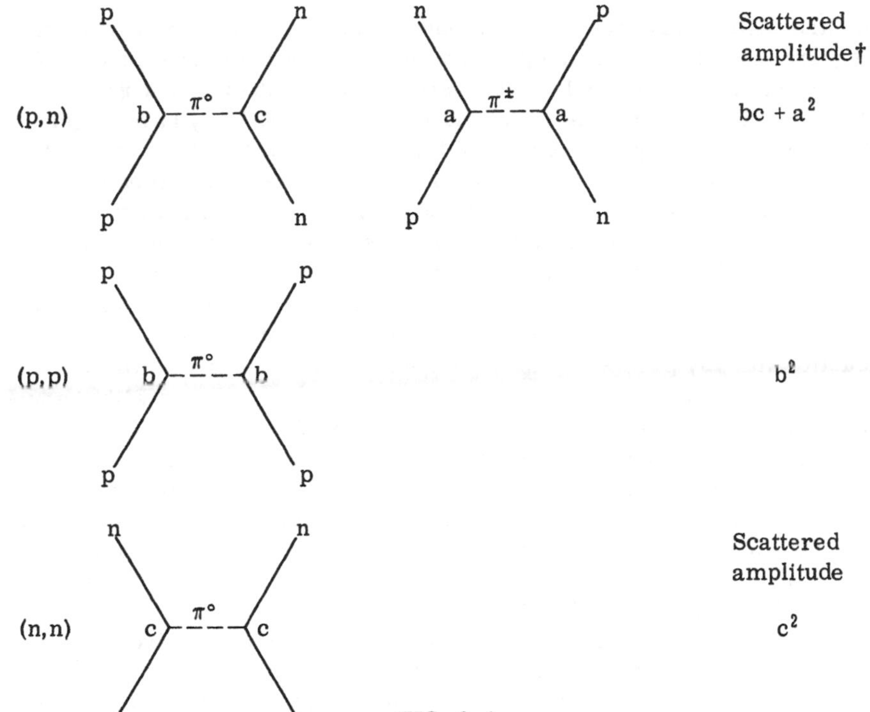

FIG. 8-1

---

†That we must add the amplitudes in the first process is somewhat tricky. To compare the (p,n) with the (p,p) and the (n,p) systems we should have considered a (p,n) state that is symmetric under the exchange of labels. We have

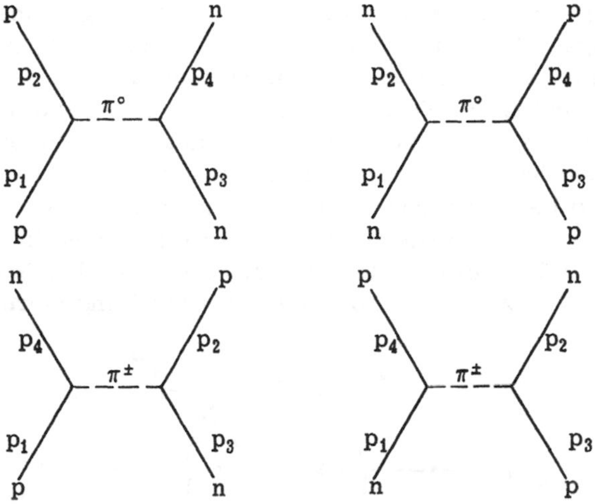

which correspond to transitions between the same initial and final states in both cases, and according to our rule the amplitudes must be added.

where the coefficients give the amplitude of each state. Probabilities are the square—so if a nucleon and pion are in a state of isotopic spin 1/2 of plus charge, the odds are 2:1 to be n and π^+ rather than p and π°.

These π mesons, or pions, have indeed been found. They have 0 spin. The π^+ and π^- have equal masses 276 times the electron, while the π° differs only slightly, 268 times the electron. The difference is probably simply the additional electrical energy of the charged pions. All the implications of isotopic spin symmetry regarding the coupling coefficients have been verified, if corrections for electrodynamics are included.

Is the nuclear force correctly given now from the exchange of pions between nucleons? This brings us to a new serious matter. We are unable to calculate, with any precision, the consequences of a strong coupling! So we cannot calculate the nuclear force directly and see if it agrees with the hypothesis of the couplings (8-1) and (8-2). We saw that there were forces resulting from the exchange of one, or of two, and of course of three or more pions. It is easy to calculate the force from one exchange, a bit harder for two, etc., but we do not know how to do the sum at all well. In electrodynamics there is also a force from the exchange of one, two, etc., photons, but the amplitude contributed by a diagram containing an extra photon is $e^2/\hbar c$ or 1/137 times as large. Thus the dominant contribution is one photon exchange, with only a few per cent correction for two photon exchange, etc. We thus work out a series of rapidly decreasing terms (called a perturbation expansion). But for the mesons the coupling constant g corresponding to e of photon coupling satisfies $g^2/\hbar c = 15$. This is very large, justifying the term "strong" coupling, but also forbidding the perturbation expansion.

A great deal has been done by invoking over-all theorems, isotopic spin symmetries, and dispersion relations (relations connected to the principle that signals cannot travel faster than light, which we cannot consider here). Suffice to say that at present it is not possible to calculate most things involving strong coupling. A serious problem is holding up the analysis of these couplings. There are even serious doubts that a strong coupling is a logically consistent possibility in quantum field theory.

Indirect Interactions. To take an example of the kind of problem involved, consider the interaction of neutrons and photons. Experimentally there is one; the neutron has a magnetic moment, known to a few parts per million. But we may still assume there is no direct neutron-photon coupling. For, by (7-4), the neutron can transform to charged particles in a virtual state, and thus indirectly interact with photons. One possibility is represented by Fig. 8-2 but there are many more diagrams involving more virtual mesons.

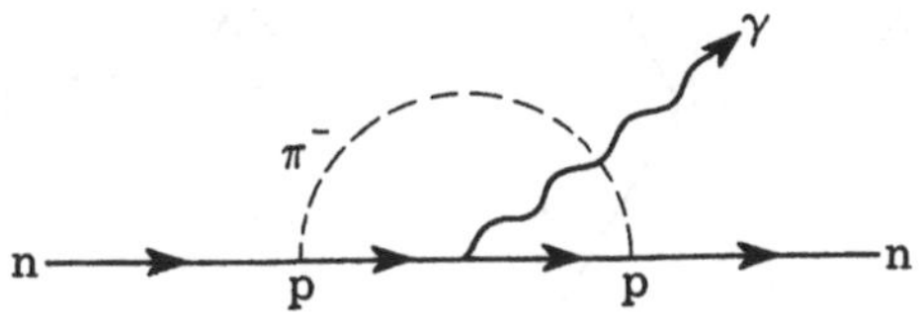

FIG. 8-2

We are unable to calculate the moment and thus cannot use the beautifully precise measurement as a test of our theories. All we can do is explain qualitatively the electrical and decay properties of pions and nucleons by such intermediate processes.

The charged and neutral pion differs markedly in their disintegration properties. The π° disintegrates very rapidly ($< 10^{-15}$ sec) into two photons,

$$\pi^\circ \rightarrow \gamma + \gamma \tag{8-3}$$

We cannot argue that isotopic spin symmetry implies an analogous reaction for the π^+. In fact, it is impossible because of conservation of charge. But (8-3) can be an electromagnetic interaction and thus isotopic spin symmetry is lost for this. We may hope to explain (8-3) as a result of the passage through a virtual pair of proton and antiproton:

$$\pi^\circ \rightarrow \bar{p} + p \rightarrow \bar{p} + p + \gamma' \rightarrow \gamma'' + \gamma'$$

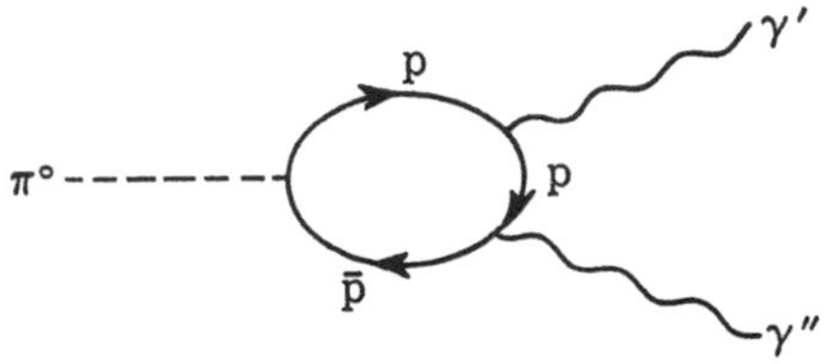

The first coupling is strong, a consequence of (7-3), next, one p emits a photon, marked γ', by (6-3), then the nucleus annihilates, emitting the second γ'' via (6-3) again. (Two photons are necessary, or momentum and energy cannot be conserved at the end. As usual energy need not be conserved in the transient intermediate stages.)

Unfortunately we cannot compute this rate either, because the first step involves strong interaction. The π^+ decays much more slowly. In 2.6×10^{-8} sec mean life it disintegrates into a μ^+ and ν,

$$\pi^+ \rightarrow \bar{\mu} + \nu \tag{8-4}$$

(As expected, the antiparticle π^- has the same lifetime for disintegration into μ and ν.)

This could be an indirect result through virtual states:

$$\pi^+ \rightarrow \bar{n} + p \rightarrow \mu^+ + \nu$$

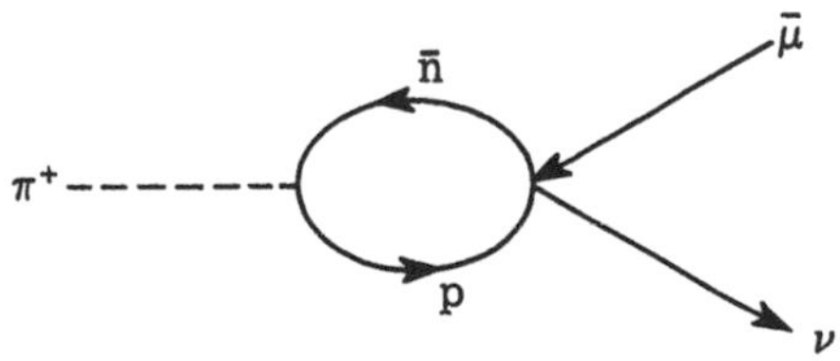

We should also expect the disintegration

$$\pi^+ \rightarrow \bar{n} + p \rightarrow \bar{e} + \nu \qquad (8\text{-}5)$$

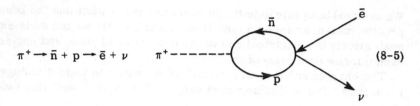

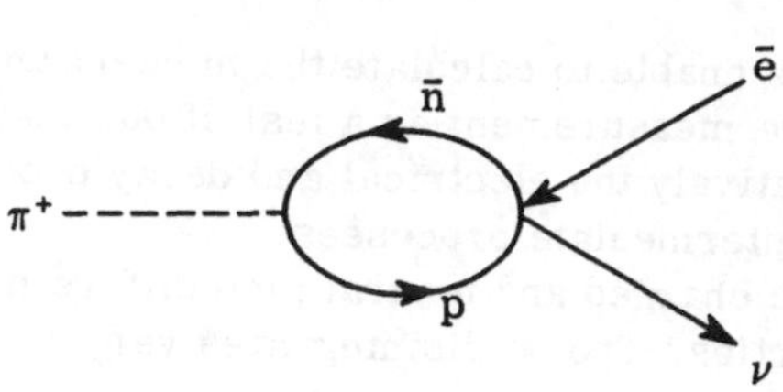

Again, because a strong coupling is involved, we cannot compute the rates directly, but we are able to compute the ratio of the rate of (8-4) and (8-5). We expect to find $\bar{e}\nu$ instead of $\bar{\mu}\nu$ as the products of pion disintegration in one case out of 7400. This has been recently confirmed experimentally (accuracy about ± 15 per cent).

9 Strange Particles

There are other particles strongly coupled to pions and nucleons. About 7 or 8 years ago some new particles were discovered in cosmic rays. For example, there was (now called Λ) a neutral particle which disintegrated into a p and π^-:

$$\Lambda \rightarrow p + \pi^- \tag{9-1}$$

It has, according to recent measurements a mass 2182 electrons and lifetime $2.6 \pm 0.2 \times 10^{-10}$ sec. This is very slow compared to the times natural for strong reactions (10^{-23} sec, the time for light to go between adjacent nucleons in the nucleus). Therefore, this decay is a weak reaction, probably related to β decay. If we limit ourselves to strong couplings, then we know there is not such a coupling as

$$\Lambda \longleftrightarrow p, \pi$$

It is forbidden as a strong interaction [otherwise (9-1) would go very fast].

But then how are the Λ produced? The cosmic rays consist of fast protons hitting nuclei that contain protons and neutrons and virtual pions via the strong coupling (6-4). The production of Λ is so copious experimentally that it must be via a strong coupling. It cannot be $p + \pi^- \rightarrow \Lambda$, for this is not strong, as we have seen. Nor can it be anything like $p + n \rightarrow \Lambda + p$, for this would imply $\Lambda \rightarrow p + n + \bar{p}$ is strong, and since $\bar{p} + n \rightarrow \pi^-$ is strong by (7-4), reaction (9-1) would be strong. Nor is a reaction such as

$$n + n + n \rightarrow \Lambda \tag{9-2}$$

possible. Because, even though (9-1) is weak, it does exist, and we would then have the possibility of three neutrons in a nucleus turning to one via the virtual reaction

$$n + n + n \rightarrow \Lambda \rightarrow p + \pi^- \rightarrow n$$

This would release a large energy, the rest mass of two neutrons, and no nucleus but hydrogen would be stable. The stability of matter like a piece of carbon is very striking; such delicate experiments to detect disintegrations have failed that we know what are commonly called stable nuclei have a lifetime of at least 10^{17} years.

This leads us to another principle which we use in choosing couplings. No couplings must exist weak or strong, so that taken all together, nucleons can disappear or disintegrate into something lighter. Thus, since one proton is produced on disintegration of the Λ, just one nucleon must be consumed in its production. The easiest way to keep track of this is to give for each particle the number of nucleons "hidden" in it; more precisely, the net number of protons that appear in its ultimate decay products (antiprotons counted as minus). This number still has no generally accepted name; nucleonic charge has been proposed. Thus the nucleonic charge of the Λ is 1, as is that of p and n. Electrons or π mesons have 0 nucleonic charge, antiprotons have a nucleonic charge of minus 1. No fundamental particles are known that have nucleonic charge greater than 1.

Then we have the principle that all couplings must satisfy the rule: *Nucleonic charge is always conserved.*

Associated Production. K Mesons. Following arguments of this kind it became clear that in the strong production more than one strange particle must be produced at once (for example, $n + n \rightarrow \Lambda + \Lambda$). Actually other particles had been discovered in cosmic rays, for example, a neutral particle, now called K° meson or neutral kaon, which disintegrates into two pions

$$K^\circ \rightarrow \pi^+ + \pi^- \tag{9-3}$$

and has a lifetime of 10^{-10} sec. It has a mass 966 and, clearly, 0 nucleonic charge. Again we encounter a slow decay, presenting the same problem as the Λ with regard to its production.

Pais and Gell-Mann suggested that they must be produced together and that the true production reaction was the result of a strong coupling

$n \longleftrightarrow \Lambda, K^\circ$

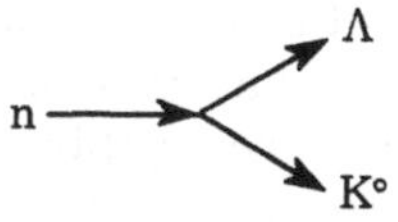

(9-4)

It has since been verified directly that these particles are produced at the time when nucleons collide.

But a strong coupling like (9-4) for the proton would affect the nuclear forces (see Fig. 9-1). The balance of n,n and p,p forces then could not be maintained unless there was an analogous coupling for the proton. There does not appear to be any charged particle analogous (that is of nearly the same mass) to the Λ so we are led to expect a coupling

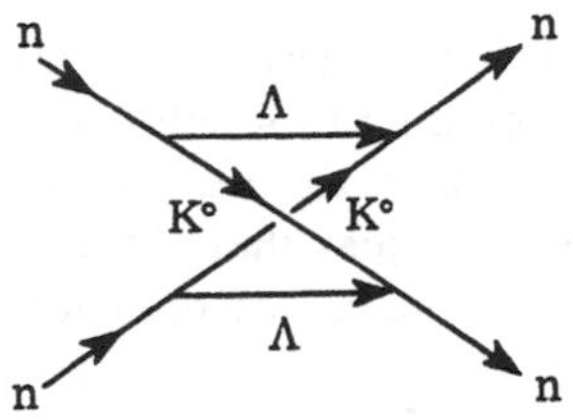

FIG. 9-1

$$p \longleftrightarrow \Lambda, K^+ \tag{9-5}$$

and there is indeed a charged counterpart to the K° found first in cosmic rays; a K^+ of mass 967. To put it in the language of isotopic spin (8-1, 8-2), the nucleon on the left side is isotopic spin 1/2, so if isotopic spin is conserved the total on the right side must be 1/2. The Λ is a singlet and so is an isotopic spin zero, so the K must be of isotopic spin 1/2, a doublet, coming in two varieties.

Although its strong coupling (9-5) is the same as the K° in (9-4) and the decay [analogous to (9-3)]

$$K^+ \rightarrow \pi^+ + \pi^0$$

does occur, the lifetime of the K^+ is 1.2×10^{-8} sec, very much longer than the K°, a reminder that the weak decays do not maintain isotopic spin symmetry.

There were still other particles discovered in cosmic rays in a confusing profusion. The clue to organizing this material was to formulate this idea, that the strong couplings involve such particles in pairs, in terms of a new principle.

Strangeness. Let us suppose that the K° is considered to carry some new kind of charge, which nucleons and pions lack, and suppose this charge cannot be created or destroyed in strong couplings. Then a single K cannot be created by strong interactions, but if the Λ is considered to carry a negative unit of this charge, its creation along with the K is allowed. This charge has been called "strangeness." The K° and K^+ have strangeness 1, the Λ has strangeness minus 1, nucleons and mesons strangeness 0. Total strangeness on both sides of a strong coupling must balance. Strangeness may change via a weak coupling [as evidenced by the existence of (9-3)].

Following these ideas Gell-Mann and Nishijima were led, independently, to propose a scheme for organizing our knowledge of strange particles and to predict many relations among them. Using this scheme, analogous to the periodic table of the chemical elements, we shall describe the strongly interacting particles believed to exist today.

Baryons. First we take the particles of nucleonic charge 1. These as a group have been called hyperons. They are illustrated in Table 9-1. The existence of the Σ° was predicted by this scheme. It was predicted that it would decay in an extremely short time to the Λ by photon emission. It was subsequently discovered and did just that. Mass near 2584, a doublet (T = 1/2) Ξ^-, Ξ°, strangeness −2. The Ξ^- or cascade particle was already known

TABLE 9-1

Isotopic spin T	Charge			Strangeness
	−	0	+	
1/2	Ξ^-	Ξ°		−2
1	Σ^-	Σ°	Σ^+	−1
0		Λ		
1/2		n	p	0

In order of their masses we find:

Mass near 1836 a doublet (T = 1/2), the nucleons n,p	0
Mass 2182 a singlet (T = 0), the Λ, neutral	−1
Mass near 2330 a triplet (T = 1), $\Sigma^-\Sigma^\circ\Sigma^+$	−1

from cosmic-ray experiments. (It has recently also been produced in the laboratory by high-energy machines.) It decays slowly (about 3×10^{-10} sec) into a Λ and π^-. This could be explained easily only if it had strangeness −2. In fact, two K mesons have been observed to be created with it. That it should be a doublet is the consequence of a relation of strangeness charge and isotopic spin suggested by Gell-Mann and Nishijima: *The strangeness S is twice the average electric charge q of each multiplet, minus the nucleonic charge N.* For hyperons then with nucleonic charge 1 it is S = 2q − 1. So the average charge of the Ξ multiplet must be −1/2 since S = −2, and thus it must be a doublet. The predicted Ξ° particle has recently been found.

Antibaryons. With each of these hyperons there should be a corresponding antiparticle of nucleonic charge −1. Thus the chart for antihyperons is exactly the same, with particles of the same mass but of opposite electrical charge and opposite strangeness. Of these, the antineutron $\bar{n}$, the antiproton $\bar{p}$, and, very recently, the antilambda $\bar{\Lambda}$ have been produced artificially in the laboratory.

Mesons. Next come the strongly coupled particles of nucleonic charge 0 (generically called mesons). They are given in Table 9-2. The π^+ and π^- are antiparticles and the π° is its own antiparticle. But since the kaons K^+, K° have strangeness + 1, their antiparticles must have strangeness −1, and in particular there must be two neutral kaons, of strangeness 1 and −1, respectively.

These are all the particles generally accepted to exist at present. There are a very few cosmic-ray events whose interpretation remains puzzling and which may be evidence of still other particles. Furthermore, cosmic-ray evidence for the existence of particles of mass near 500 electrons has been found by one laboratory but other attempts to find these particles have failed so far. The existence of all but some of the antihyperons is well established experimentally.

Problem 9-1: An experiment has been done on K capture in deuterium: $K + D \rightarrow$ hyperon + pion + nucleon. The data given in Table 9-3 are available. At present one cannot distinguish the Λ° from the Σ°

Can you test the principle of conservation of isotopic spin?

Make as many predictions as you can, in particular for the results if Λ° and Σ° could be distinguished.

TABLE 9-2

Isotopic spin T	Charge			Strangeness
	−	0	+	
1/2	K^-	K°		−1
1/2		K°	K^+	+1
1	π^-	π°	π^+	0

In order of mass we find:

Mass near 276 a triplet (T = 1) the pions π^- π° π^+	0
Mass near 965 a doublet (T = 1/2) the kaons K°, K^+	+1
and their antiparticles (T = 1/2) the kaons $\bar{K}^\circ$, K^-	−1

TABLE 9-3

Products	No. of cases observed
$\Sigma^+ N \pi^-$	44
$\Sigma^- N \pi^+$	55
$\Sigma^- p \pi^\circ$	7
$\Lambda^\circ p \pi^-$ $\Sigma^\circ p \pi^-$	48 (total)
$\Lambda^\circ N \pi^\circ$ $\Sigma^\circ N \pi^\circ$	72 (total)

10 Some Consequences of Strangeness

The concept of strangeness and its conservation in strong interactions has led to a large number of predictions, none of which have been violated by experience. It has served very faithfully to help organize the experimental material. These predictions are, for example, that when a Λ or Σ is produced in nuclear collisions, a K° or K^+ must be also. Again, a production reaction such as $n + n \rightarrow \Lambda + \Lambda$ is impossible for the total strangeness of the two Λ's is -2, and of the neutrons 0.

As another example, K^- particles colliding with nuclei in flight may produce Λ, but K^+ cannot do so.

The Decay of The Neutral Kaon. One of the most strikingly brilliant predictions of this theory of strangeness was made by Pais and Gell-Mann. It is related to the prediction that there must be two neutral K particles, having opposite strangeness, K°, and its antiparticle $\bar{K}^\circ$. Now the K° appears to decay into two pions, for example, the disintegration

$$K^\circ \rightarrow \pi^+ + \pi^- \tag{9-3}$$

is observed (lifetime about 10^{-10} sec). This violates strangeness, of course, as weak interactions do. The antiparticle should decay with the same probability into the corresponding antiparticles

$$\bar{K}^\circ \rightarrow \pi^- + \pi^+ \tag{10-1}$$

The products in (9-3) and (10-1) are of course identical. This has, as a consequence, a very interesting quantum mechanical interference effect. The existence of (9-3) means that, even though it may be via complex virtual processes, there is some amplitude, x say, for a K° to become π^+,π^-:

$$K^\circ \longleftrightarrow \pi^+,\pi^- \qquad \text{amplitude } x \tag{10-2}$$

Also, from the relation of particle and antiparticle, the amplitude for the antiparticle must be the same,

$$\bar{K}^\circ \longleftrightarrow \pi^+,\pi^- \qquad \text{amplitude } x \tag{10-3}$$

(Strictly it might have opposite sign, but either possibility leads to the same conclusion.)

Now suppose we had gotten the particle in a state neither K° nor $\bar{K}^\circ$ but rather with equal but opposite amplitudes to be K° and $\bar{K}^\circ$; call the state K_2°:

$$K_2^\circ = (1/\sqrt{2})K^\circ - (1/\sqrt{2})\bar{K}^\circ \tag{10-4}$$

(The amplitude must be $1/\sqrt{2}$ because the probability (amplitude squared) is 1/2 to be K° and 1/2 to be $\bar{K}^\circ$.) Then the particle in this state K_2° could not decay into π^+ and π^-, for the amplitude for it to do so is $[(1/\sqrt{2})x - (1/\sqrt{2})x] = 0$, from (10-2) and (10-3)! The state

$$K_1^\circ = (1/\sqrt{2})K^\circ + (1/\sqrt{2})\bar{K}^\circ \tag{10-5}$$

could, of course, decay with amplitude $\sqrt{2}\,x$.

Thus the proper states to use to describe disintegration are K_1° and K_2°, for the first may, and the second may not, decay into two pions. These two states will have very different lifetimes and disintegration products. (It turns out the K_2° can disintegrate, for example, into three particles, and has a lifetime at least 100 times as long as the K_1° going to two pions.) But when a kaon is produced, say along with a Λ, it has a definite strangeness -1; it is K°, and is neither K_1° nor K_2°:

$$K^\circ = (1/\sqrt{2})K_1^\circ + (1/\sqrt{2})K_2^\circ \tag{10-6}$$

an equation immediately deduced by adding (10-4) and (10-5).

The probabilities are thus 1/2 to be K_1° and 1/2 to be K_2°, the proper states for analysis of disintegration. Thus when the newly produced kaons disintegrate, only half of them should exhibit the short lifetime 10^{-10} sec and decay into two pions. The remainder should have a much longer lifetime, 10^{-8} sec, and decay into three particles. That is, the remarkable prediction was made that the neutral kaon should exhibit two different lifetimes, with two sets of decay products. This has now been verified. A further aspect has also been verified. Since the K° has strangeness +1, it cannot produce Λ in collision with nuclei. However, if we move far enough away from the source of a K° beam so that K_1° would nearly certainly have disintegrated (but not so far that a K_2° would also disintegrate appreciably) the beam must become nearly purely in the state K_2°. But by (10-4) the strangeness is not definite now; there is an amplitude $-1/\sqrt{2}$, and thus a probability 1/2 to be $\bar{K}^\circ$ of strangeness -1. Then Λ's can be produced by the kaon beam hitting nuclei. This Λ production has now been demonstrated experimentally.

It may be asked, in an over-all way: How did the strangeness change from +1 on production to -1 further down in the beam? The answer is via the virtual process (10-3) followed by (10-3) in reverse. The strangeness violation is via a very weak coupling x^2, it is true, but the equality of mass of K° and $\bar{K}^\circ$ makes a resonance possible, so that even the small amplitude for the process gradually builds up a large effect.

This is one of the greatest achievements of theoretical physics. It is not based on an elegant mathematical hocus-pocus such as the general theory of relativity yet the predictions are just as important as, say, the prediction of positrons. Especially interesting is the fact that we have taken the principle of superposition to its ultimately logical conclusion. Bohm and co-workers thought that the principles of quantum mechanics were only temporary and would eventually fail to explain new phenomena. But it works. It does not prove it right, but for my money, the principle of superposition is here to stay!

11 Strong Coupling Schemes

Given now the particles, hyperons, and mesons, and knowing that they are strongly coupled, the next problem is to find which are strongly coupled to which and in what manner. These couplings must satisfy conservation of nucleon number, charge, isotopic spin, and strangeness, but that is far from determining them completely. For example, what coupling does the π have besides the nucleon coupling (8-1), (8-2)? Is there a coupling $\Lambda \rightarrow \Sigma,\pi$ and of what strength and kind, etc.? Again what is the law of coupling of the K's in (9-4) and in reactions such as $N \longleftrightarrow \Sigma,K$ or $\Sigma \longleftrightarrow \Xi,K$? Even the spins of the particles are uncertain. Studies of the decay $K^+ \rightarrow \pi^+ + \pi^+ + \pi^-$ indicate strongly that no angular momentum is carried out by the pions, so the kaon spin is probably 0. But the experimental evidence is incomplete on Λ, Σ, and Ξ, although spin 1/2 is indicated for the Λ and Σ. Probably the hyperons all have spin 1/2, the mesons spin 0.

Fermi and Yang Model. As an example of a host of models proposed to understand the strong interactions, I shall discuss an idea of Fermi and Yang.

Suppose a neutron and a proton have a charge analogous to electrical charge but of the same sign, which couples them to a vector meson of very large mass. Then the $n\bar{p}$ system would feel a short-range attractive force analogous to the longer-range electrical attraction $e\bar{e}$. The n and $\bar{p}$ have together a mass of 2×938 Mev and we imagine that they are bound together very strongly, say, by about 1600 Mev. If the forces are right, the total angular momentum would be 0, the parity −1, and we obtain the π^-: $[n\bar{p}] = \pi^-$. Also, $[p\bar{n}] = \pi^+$. Typical diagrams are of the form shown in Fig. 11-1, where the wiggly line represents vector-meson exchange. We also expect a $p\bar{p}$ and $\bar{n}n$ system. Which of these is the π°?

We note that the $p\bar{p}$ and $n\bar{n}$ systems have additional diagrams (Fig. 11-2) which do not occur in the $n\bar{p}$ and the $p\bar{n}$ cases. It turns out that the system†

†Because the $\bar{p}p$ and $\bar{n}n$ both have the same amplitude to annihilate, so the state $(1/\sqrt{2})(\bar{p}p - \bar{n}n)$ cannot annihilate. In judging whether it is an isotopic spin triplet, one must remember to count $\bar{n}$ as $-(1/2)$.

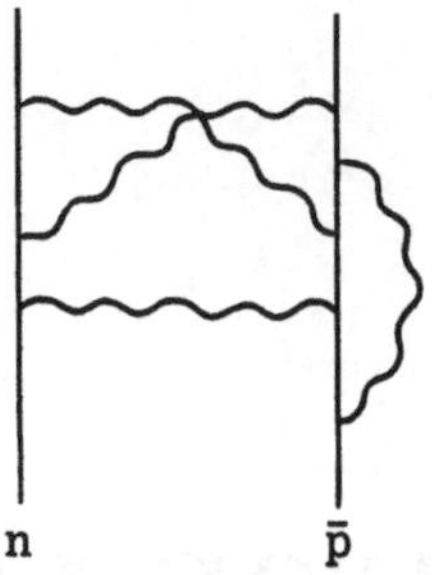

FIG. 11-1

which has the same energy as a π^+ and a π^- and does not annihilate is

$$(1/\sqrt{2})([\bar{p}p - \bar{n}n]) = \pi^\circ$$

These form an isotopic spin triplet. The other state,

$$(1/\sqrt{2})([\bar{p}p] + [\bar{n}n]) = \Delta^\circ$$

might not be bound or might have a different energy, perhaps higher, and be a new meson of total isotopic spin T = 0, which has not yet been observed.

To get the remaining particles it is necessary to introduce at least one

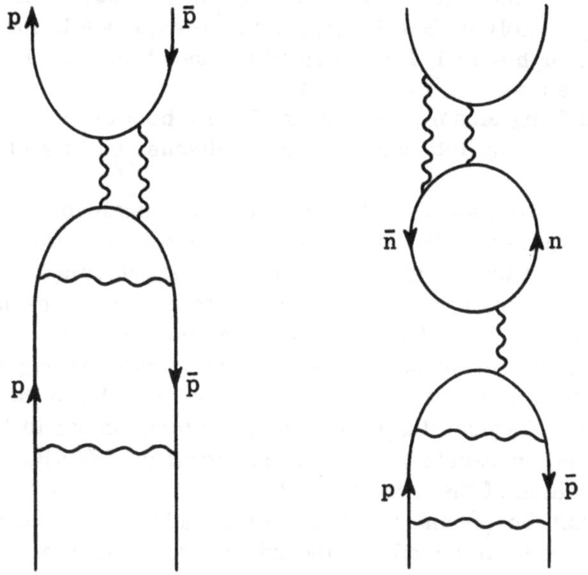

FIG. 11-2

"fundamental" particle which carries strangeness. Choose the Λ. Then

$$(\Lambda\bar{p}) = \bar{K}$$

$$(\bar{\Lambda}p) = K^+$$

$$(\Lambda\pi^{+,\circ,-}) = \Sigma^{+,\circ,-}$$

$$(\Lambda\Lambda\bar{p}) = \Xi^- \qquad (\Lambda\Lambda\bar{n}) = \Xi^\circ$$

Strangeness is just the number of Λ's! So you see, it is possible to imagine all the strongly interacting particles as composites of n, p, and Λ and deduce isotopic spin and strangeness conservation.

I shall tell you my secret hypothesis: You cannot tell whether a particle is "elementary" or whether it is a composite of "elementary" particles. In other words, all the theories of composite particles would give equivalent results (if we could calculate them) and there would be no way to distinguish among these.

If you have a system composed of particles of mass large compared to the total binding energy (nuclei, atoms) then it makes sense to speak of a composite system. But when the binding energy is a good fraction of the mass of the free particles it is incorrect to make a distinction between composite and elementary particles. How can this idea be more clearly stated and how can practical use be made of it? I don't know.

The theoretical proposals for the detailed scheme of strong couplings are all nearly entirely speculative. We shall give two others; one is the Gell-Mann[7] suggestion, called global symmetry. It proposes that all the hyperons would have the same mass and be various states of an eightfold multiplet if it were not for the K couplings. Two new states are formed by linear combinations of Λ and Σ°; namely, $Y = (1/\sqrt{2})(\Lambda - \Sigma^\circ)$ and $Z = (1/\sqrt{2})(\Lambda + \Sigma^\circ)$. Then it is assumed that the form and size of the pion couplings are unchanged if, in the coupling, (8-1), (8-2) of the pions with n and p, the n,p are replaced by Y, Σ^+, or are replaced by Σ^-, Z, or are replaced by Ξ^-, Ξ°. The kaon couplings that destroy this symmetry remain undetermined.

The other suggestion is that the pions are coupled directly to the total isotopic spin vector, and this coupling is responsible for the Λ, Σ splitting. The kaons couple between n, p and the Λ,Σ quartet, and between Ξ^-, Ξ° and the Λ, Σ quartet in such a way as not to split the Λ, Σ quartet (but the coupling to n,p and to Ξ^-, Ξ° are with different coefficients). (In the notation of Gell-Mann,[7] we take $g_{\Lambda\pi} = 0$, $g_{N\pi} = 1/2$; $g_{\Sigma\pi} = g_{\Xi\pi}$, $g_{\Sigma K} = g_{\Lambda K}$; and $h_{\Sigma K} = h_{\Lambda K}$.) The pattern of mass values expected in this scheme fits very well to that observed. Unfortunately critical tests of this hypothesis have not yet been found.

There is a large experimental program on to determine production of kaons by nuclear collisions and by photons, scattering and interaction of these mesons with nuclei, etc. But just between us theoretical physicists: What do we do with all these data? We can't do anything. We are facing a very serious problem and we need a revolutionary idea; something like Einstein's theory. Perhaps the results of all experiments will produce some idiotic surprises, and someone will be able to calculate everything from some simple rule. What we are doing can be compared with those complicated models invented to explain the hydrogen spectra which turned out to satisfy very simple regularities.

One more thing about the question of strong couplings. There is also direct evidence that the Λ interacts strongly with nucleons. There exist hyperfragments (a better name would be hypernuclei), in which a Λ is bound to a number of nucleons. For example, the hypernucleus ${}_{\Lambda}He^4$ has been found as a fragment resulting from the capture of a K^- by a nucleus. This hypernucleus consists of two protons, a neutron, and a Λ bound together. The Λ is bound by a few Mev. The system is unstable of course, for the weak decay (9-1) of the Λ provides a mechanism for disappearance of the Λ with release of a pion and 37 Mev (or the pion may be virtual or recaptured and its rest energy appears as kinetic energy of nucleons in a star). From a study of these hypernuclei we may eventually get information on the Λ-nucleon interaction force. At any rate, it is nearly as strong as the nucleon-nucleon interaction. For further details see the review article by Dalitz.[8]

Additional evidence on these strong couplings should eventually come from a study of their relation to weak couplings. For example, magnetic moments and electromagnetic mass differences, as well as the relative rates of various weak decay processes must tell us something about the structure of the strongly interacting particles. Yet the theoretical analysis of all this evidence related to strong couplings is severely crippled by the inability to make quantitative calculations with such couplings.

12 Decay of Strange Particles

We next turn to the evidence on the weak decay of these particles. The experimental information on masses and decay properties of all the particles is given in Table 12-1. We are concerned here only with the decays of hyperons and mesons. Two of the decays clearly are the result of electromagnetic couplings (in association with virtual states implied by strong cou-

TABLE 12-1

Products	Ratio, %	Lifetime, sec
$n \to p + e + \bar{\nu}$		1040
$\Lambda \to p + \pi^-$	63 ± 3	2.6×10^{-10}
$\quad n + \pi^\circ$	37 ± 3	
$\Sigma^+ \to p + \pi^\circ$	46 ± 6	0.8×10^{-10}
$\quad n + \pi^+$	54 ± 6	
$\Sigma^- \to n + \pi^-$	100	1.6×10^{-10}
$\Xi^- \to \Lambda + \pi^-$	?	$\approx 10^{-9}$
$\pi^+ \to \mu^+ + \nu$	100	2.56×10^{-8}
$\quad e^+ + \nu$	0.013	
$K^+ \to \mu^+ + \nu$	59 ± 2	1.22×10^{-8}
$\quad \pi^+ + \pi^\circ$	26 ± 2	
$\quad \pi^+ + \pi^+ + \pi^-$	5.7 ± 0.3	
$\quad \pi^+ + \pi^\circ + \pi^\circ$	1.7 ± 0.3	
$\quad \pi^\circ + e^+ + \nu$	4.2 ± 0.4	
$\quad \pi^\circ + \mu^+ + \nu$	4.0 ± 0.8	
$K_1^\circ \to \pi^+ + \pi^-$	78 ± 6	1.0×10^{-10}
$\quad \pi^\circ + \pi^\circ$	22 ± 6	

pling), the $\pi^\circ \to \gamma + \gamma$ and the $\Sigma^\circ \to \Lambda + \gamma$. These are also the only ones allowed by conservation of charge and the principle that electromagnetic couplings cannot alter strangeness.

The remaining decays all have lifetimes of the same general order of magnitude. It is believed that they are all the result of coupling of the Fermi type (in association with virtual states implied by strong couplings, as always). This hypothesis would already account for several features including the general order of magnitude of lifetimes. Certainly lepton emissions would not be unexpected. But even when no leptons are involved a lack of conservation of parity (an asymmetry under reflection of left for right) speaks for a Fermi coupling. The fact that the same particle K could decay into two and into three pions (of 0 total angular momentum) was in fact the first suggestion that conservations of parity may be violated in physical law. More recently an asymmetry demonstrating a failure of reflection symmetry has been found in the direction of the products in the decay of the Λ to $p + \pi^-$. We have already written three Fermi couplings (6-4), (6-5), and (6-6), but in each case, counting the leptons as of 0 strangeness, no strangeness change is implied. Thus with these three, we can only hope to explain decays for which $\Delta S = 0$ where S is total strangeness. They are only the neutron decay $n \rightarrow p + e + \bar{\nu}$ and the pion decay $\pi^+ \rightarrow \mu^+ + \nu$, which we have already discussed as an indirect consequence of (6-6) via virtual states.

The remaining decays involve a change of strangeness of one unit, $\Delta S = \pm 1$. The fundamental couplings that produce them have not been identified, although they are almost certainly Fermi couplings. The problem of their identification is interesting, so we shall go into it in some detail.

Fermi Coupling Schemes. We shall first count the minimum number of new couplings needed. First of all, the existence of a decay like $K^+ \rightarrow e^+ + \nu + \pi^\circ$ implies that the $\bar{\nu}e$ are coupled to a strangeness changing pair. It need not be $\bar{K}^+\pi^\circ$ directly, as the strong couplings would allow such a decay if any other pair, such as $\bar{p}\Lambda$ were coupled to $\bar{\nu}e$. For example (Fig. 12-1), the K^+ could become virtually $\bar{\Lambda}$ and p, and the $\bar{\Lambda}$ decay to $\bar{p}\bar{e}\nu$

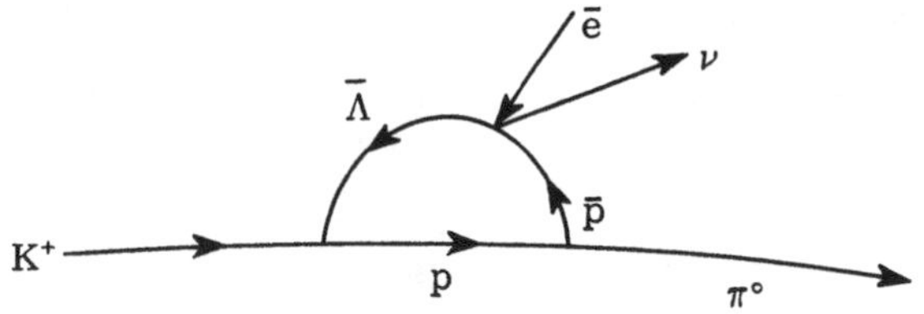

FIG. 12-1

with the p and $\bar{p}$ annihilating to π°. Let us take, as an example, a Fermi coupling

$$\bar{p}\Lambda \longleftrightarrow \bar{\nu}e \qquad (12\text{-}1)$$

analogous to (6-4). But muons are emitted also in $\Delta S = -1$ decays (e.g., $K^+ \rightarrow \mu^+ + \nu$ or $K^+ \rightarrow \mu^+ + \nu + \pi^\circ$), so we must have an additional coupling

$$\bar{p}\Lambda \longleftrightarrow \bar{\nu}\mu \tag{12-2}$$

Finally there are decays involving no neutrinos at all. These could come from a Fermi coupling of the type

$$\bar{p}\Lambda \longleftrightarrow \bar{p}n \tag{12-3}$$

of any equivalent via strong couplings. For example, the $\Lambda \rightarrow p + \pi^-$ decay would be via the virtual process, such as

$$\Lambda \xrightarrow[(12-3)]{} p + \bar{p} + n \xrightarrow[(7-4)]{} p + \pi^-$$

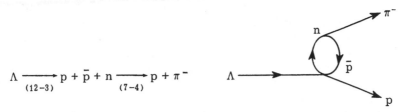

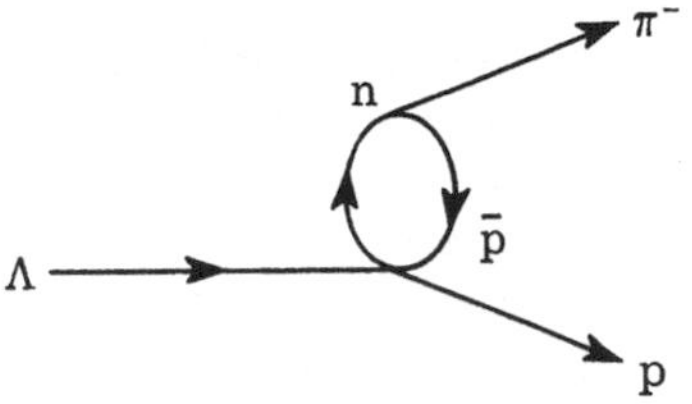

plus more complicated diagrams. With these three couplings all known decays would be qualitatively explainable.

The proposal of six independent Fermi couplings might appear complicated, but the simultaneous appearance of the three new ones will appear natural from one assumption. That is that the Fermi couplings are of the nature of the interaction of a kind of current with itself:

$$J \longleftrightarrow J \tag{12-4}$$

and the problem is to find the composition of the current J, the sum of several parts. The couplings (6-4), (6-5), and (6-6) described previously result if J is written

$$J = (\bar{\nu}e) + (\bar{\nu}\mu) + (\bar{p}n) + X \tag{12-5}$$

Experimentally the coefficient of all first three terms are equal. All our three new couplings will result if we add to J just one term, say X, which changes strangeness. Above we have suggested solely as an example what X might be but we shall now have to consider more seriously what properties the term X might have.

An immediate consequence of this idea is that the coefficients of X to each of the three currents $(\bar{\nu}e)$, $(\bar{\nu}\mu)$, and $(\bar{p}n)$ are equal. That is, the couplings (12-1), (12-2), and (12-3) must all have the same coefficient [although it need not equal the coefficient of (6-4), (6-5), and (6-6)].

If the couplings of $(\bar{\nu}e)$ and $(\bar{\nu}\mu)$ to X are equal, it can be calculated that about as many K^+ decays go to $e^+ + \nu + \pi^\circ$ as go to $\mu^+ + \nu + \pi^\circ$. This is supported by the data to an accuracy of 30 per cent. The absence found so far of $K^+ \to e^+ + \nu$ is also consistent if the K^+ has spin 0, for then the rate of $K^+ \to e^+ + \nu$ should be very much less than the rate $K^+ \to \mu^+ + \nu$.

Can we check the prediction that the couplings of $(\bar{p}n)$ and of $(\bar{\nu}\mu)$ to X are equal? Unfortunately, because of our inability to analyze strong couplings, we have found no way to do it yet.

Proposed Symmetry Rules of the Strangeness Changing Decays. What can we say about the current X? We try to make as restrictive hypotheses as possible; in that way we make the maximum of predictions, although some of them may be proved wrong in the future. First we notice that the presence of $K^+ \to e^+ + \nu + \pi^\circ$ requires that, if strangeness decreases, positrons are emitted. So X must have at least one term like $(\bar{p}\Lambda)$, consisting of a pair of particles of total strangeness -1 (and in any case of negative charge). Can there be also a term in X like $(\bar{\Sigma}^+ n)$ of strangeness $+1$? There is no evidence that there must be, so we shall assume:

1. *X contains only terms of strangeness −1.* This has as a consequence that a decay with $\Delta S = 2$ is forbidden. Thus the Ξ^- could not decay into $n + \pi^-$ and so far it has not been seen to do so.

Furthermore, although a K_2° could disintegrate equally into $\bar{e} + \pi^- + \nu$ and $e + \pi^+ + \bar{\nu}$, only the first is allowed for K° and the second for $\bar{K}^\circ$. So, if the rare lepton decays of a K° beam be observed close to the course (before the K_1° mode has decayed away), the charged leptons (e or μ) should be predominantly positive. Experiments to verify this have not yet been performed.

Next, the current X, if $(\bar{p}\Lambda)$ would be of isotopic spin 1/2; other combinations, such as $(\bar{p}\Sigma^\circ)$ alone, would bring in some isotopic spin 3/2. If it were purely isotopic spin 3/2, there would be a rule, $\Delta T = 3/2$, in a leptic decay. But the decay $K^+ \to \mu^+ + \nu$ of the K^+ of isotopic spin 1/2 would then be impossible (for here $\Delta T = 1/2$). So X must contain at least some component of isotopic spin 1/2. We shall assume it is pure isotopic spin 1/2 [so if $(\bar{p}\Sigma^\circ)$ appears it must be in the combination $-(\bar{p}\Sigma^\circ) + \sqrt{2}(\bar{n}\Sigma^-)$]. Then we would have the rule:

2. *In a leptic decay the isotopic spin change can only be 1/2.* We can test this rule by comparing the decays

$$K^+ \to \pi^\circ + e^+ + \nu \tag{12-6}$$

and

$$K^\circ \to \pi^- + e^+ + \nu \tag{12-7}$$

What the rule says is that if the K's are brought to the other side to make antiK's the pairs $(\bar{K}^+\pi^\circ)$ and $(\bar{K}^\circ\pi^-)$ must have amplitudes in the proportions $-(\bar{K}^+\pi^\circ) + \sqrt{2}(\bar{K}^\circ\pi^-)$ to make up a state of isotopic spin 1/2 [analogous to expression (8-2)]. Thus the amplitude for the second reaction (12-7) is $-\sqrt{2}$ times the first (12-6), so the rate of the K° decay is twice that of the K^+. The decay of K° must be

$$\bar{K}^\circ \rightarrow \pi^+ + e^- + \bar{\nu} \tag{12-8}$$

the antiparticle reaction corresponding to (12-7) and it must occur at the same rate as (12-7) [i.e., twice the rate of (12-8)]. The K_2° particle, being of amplitude $1/\sqrt{2}$ to be K° and $-1/\sqrt{2}$ to be $\bar{K}^\circ$ will therefore decay into $\pi^- + e^+ + \nu$ at the rate of (12-8) and into $\pi^+ + e^- + \nu$ at the same rate. A corresponding relation applies to the decays with a muon replacing the electron. These predictions are confirmed by experiment (so far).

13 The Question of a Universal Coupling Coefficient

Nonleptic decays occur from the combination of X with the term ($\bar{p}n$) in the current J. This term ($\bar{p}n$) is of isotopic spin 1, so if combined with an X assumed to be pure isotopic spin 1/2 we can form only isotopic spin 1/2 and 3/2, and suggest the third rule governing strangeness changing decays:

3. *In nonleptic decays the change of isotopic spin can be only*

$$\Delta T = 1/2 \qquad \text{or} \qquad \Delta T = 3/2$$

This does not appear very restrictive, yet it does have consequences that we can check.

First, we can predict charge ratios for the three pion decays of the kaons. The three pions in the decay

$$K^+ \rightarrow \pi^+ + \pi^+ + \pi^- \tag{13-1}$$

seem from their momentum distribution to be in a state of zero angular momentum, and therefore the wave function of the pions is completely symmetrical. It can be shown that the only symmetrical isotopic spin states available to three particles, each of isotopic spin 1, are $T = 1$ and $T = 3$. If rule 3 is correct, the state $T = 3$ cannot be generated from the original $T = 1/2$ of the kaon, for a change of at least 5/2 would be involved. Thus the final state must be $T = 1$, and from the rules for combining states it is easy to show that the rate of the decay $K^+ \rightarrow \pi^+ + \pi^\circ + \pi^\circ$ should be 1/2 of the rate of (13-1)(except for a rate increase of about 9 per cent for each π° resulting from the small mass difference of π^+ and π°). Experimentally the ratio is 0.30 ± 0.06, which is consistent with the predicted $0.25(1.2) = 0.30$.

By exactly the same argument, the three pion decays of the K_2°,

$$K_2^\circ \rightarrow \pi^+ + \pi^- + \pi^\circ \tag{13-2}$$

$$K_2^\circ \rightarrow \pi^\circ + \pi^\circ + \pi^\circ \tag{13-3}$$

should occur in ratio 2/3 [or corrected for π° mass difference, ratio $2(1.1)/3(1.3) = 0.56$] if the final state is $T = 1$. Measurements on the K_2° are just beginning to be made, and are so far in agreement with this prediction.

There is also a consequence for the two-pion disintegration of the kaon. The data are

$K_1^\circ \rightarrow \pi^+ + \pi^-$		78 ± 6 per cent
$\rightarrow \pi^\circ + \pi^\circ$		22 ± 6 per cent
$K^+ \rightarrow \pi^+ + \pi$		0.002 times K_1° decay rate

A remarkable feature is that the K^+ decay is so much less rapid than the K_1° decay.

For two pions in a symmetrical state, the total isotopic spin is $T = 0$ or $T = 2$. Only the $T = 2$ state is available for the case K^+ of one π^+ and one π°. Now this could be reached in general from the kaon $T = 1/2$ either by $\Delta T = 3/2$, or by $\Delta T = 5/2$ (we add isotopic spin as vectors). According to hypothesis 3, however, only the $\Delta T = 3/2$ operates. That means that the rate of K^+ decay gives us the relative amplitude of $T = 2$ in the decay of K°. Actually it gives us only the square, but we know that the amplitude is 0.052 times some complex phase for the two pions of K_1° to be in state $T = 2$. This amplitude is so small that the K_1° should decay almost purely into $T = 0$. If it did so, the relative proportions of $\pi^+ + \pi^-$ to $\pi^\circ + \pi^\circ$ should be 2:1 or 67 per cent should be charged. If the amplitude for $T = 2$ has its phase for maximum or for minimum interference, the percentage predictions are 72 and 62 per cent, respectively. Theoretically then, the proportion of $\pi^+ + \pi^-$ in the K_1° must lie somewhere between 62 and 72 per cent if hypothesis 3 is valid. We must wait for more precise data to see if this is true; the present results are just consistent with it. As far as we know, then, the X current can be restricted to one whose strangeness is −1 and isotopic spin is 1/2. In terms of the known particles then, it would have the form

$$X = \alpha(\bar{p}\Lambda) + \beta[-(\bar{p}\Sigma^\circ) + \sqrt{2}(\bar{n}\Sigma^-)] + \gamma[-(\bar{K}^+\pi^\circ) + \sqrt{2}(K^\circ\pi^-)]$$

$$+ \delta[-(\bar{\Sigma}^\circ\Xi^-) + \sqrt{2}(\bar{\Sigma}^+\Xi^\circ)] + \varepsilon(\bar{\Lambda}\Xi^-)$$

with the coefficients α, β, γ, δ, and ε to be determined. This is as far as we have been able to proceed. The difficulties in proceeding further will now be pointed out.

The Question of a Universal Coupling Coefficient. In view of the apparent equality of the coefficients in J of such different terms as $(\bar{\nu}e)$, $(\bar{\nu}\mu)$, and $(\bar{p}n)$, it is natural to suggest a kind of universality and propose that the coupling coefficients of all particles are equal (universal), so that all the coefficients α to ε are equal and equal to unity. [Or at least if factors $\sqrt{2}$ are differently distributed, some of them may be 1 and others $1/\sqrt{2}$ to provide some special symmetry. For example, if $\alpha = 1/\sqrt{2} = -\beta$ the first two terms become $(\bar{p}Z) + (\bar{n}\Sigma^-)$, a combination especially simple from the viewpoint of global symmetry.]

As a further example in the second strong coupling scheme suggested above, a particularly simple hypothesis would be that (1) the nonstrangeness changing Fermi current is coupled to the same combination of particles as is the π^+, namely, a component of isotopic spin, and also (2) the strangeness changing Fermi current is coupled to the same combination of particles, as is the K^+. This would mean $\alpha = -\beta$, $\delta = \varepsilon$, and possibly $\gamma = 0$.

But there is direct evidence that this is not the case. If $\gamma = 1$, disregarding the other terms, the decay $K^+ \rightarrow \pi^\circ + e^+ + \nu$ could be a direct process, and its rate calculated. It comes out 170 times too fast! There may be some modification from other diagrams but it surely cannot be so drastically reduced. We deduce that either γ is 0 or of the order of 0.08 (i.e., $1/\sqrt{170}$, for the rate goes as γ^2). If γ were 0 the process would have to be an indirect one, which we cannot calculate (although at first guess it is not easy to see how it could be so slow even if γ were 0 and the other constants were of order unity—yet no firm conclusion can be drawn this way about the other constants). Again, if $\alpha = 1$, we can calculate a rate for the process $\Lambda \rightarrow p + e + \bar{\nu}$. This process and the one with μ replacing e have not been seen, yet we predict it should appear in 16 per cent of the Λ disintegrations. Experimentally it occurs at least less than one-tenth as often as this. It is unlikely that this is an effect of interference from other diagrams, so α is probably less than 0.3. In addition no lepton decay of the Σ^- has been seen again less than 10 per cent of that expected with $\beta = 1/\sqrt{2}$, so $\beta/\sqrt{2}$ must also be less than 0.3. It therefore does not look as if the X is coupled to leptons with the full coefficient expected from the universal coupling; in fact, a coefficient of order 0.1 seems more likely. (It is not possible to disprove this from the rapidity of the $K^+ \rightarrow \mu^+ + \nu$ decay, again because of uncertainties in all such quantitative calculations.)

We can summarize these points by the observation that, although theoretically unexpected, the data may indicate that

4. *Leptic decays with change of strangeness are relatively much slower than those without change of strangeness* (although the $K^+ \rightarrow \mu^+ + \nu$ is a possible violation).

But if the coefficients in X are of the order of 0.1 for lepton coupling, we should expect them to be exactly the same for the $(\bar{p}n)$ coupling. This is uncomfortable because the nonleptic decays seem too fast for this. They seem to require coefficients of order unity, but we cannot be sure, for we cannot really calculate these processes because of the virtual states of strongly interacting particles that are involved.

In addition there is a further approximate symmetry rule suggested by the data for which we have no theoretical explanation:

5. *Nonleptic strangeness changing decays with* $\Delta T = 3/2$ *are relatively much slower than those with* $\Delta T = 1/2$, i.e., *weaker.*

14 Rules of Strangeness Changing Decays: Experiments

We have already noticed this for the K $\rightarrow \pi + \pi$ decays, in which the neutral kaon ($\Delta T = 1/2$) decayed 500 times faster than the charged kaon ($\Delta T = 3/2$). The amplitude for $\Delta T = 3/2$ here was only 0.052 of that for $\Delta T = 1/2$.

Let us ask if a similar predominance exists elsewhere. It is best studied by seeing to what extent the other data could satisfy the rule that the nonleptic decays were entirely $\Delta T = 1/2$.

First, in the Λ decay coming from $T = 0$, the final state would have to be $T = 1/2$, and the ratio of $p + \pi^-$ cases to $n + \pi^\circ$ would have to be 2:1, or charged products in 67 per cent of the decays. The data shows 63 ± 3 per cent, a discrepancy that may either be experimental error or the result of a very small interference with $T = 3/2$.

Second, there are some predictions about the Σ-decay asymmetries, but for the present incomplete data they only represent inequalities which are indeed satisfied by the data.

Third, we can now determine the rate of K_2° decays to three pions, (13-2) and (13-3), relative to that of the K^+ (13-1), for we must reach the $T = 1$ state in a unique way if $\Delta T = 1/2$. The prediction is that the total rate of K_2° to three pions equals the total rate of K^+ to three pions (if the corrections of 9 per cent for each π° are allowed for). The preliminary measurements on the K_2° are not in disagreement with this.

We would predict that the $\Xi^\circ \rightarrow \Lambda + \pi^\circ$ rate should be half as fast as the $\Xi^- \rightarrow \Lambda + \pi^-$ rate, but data are not available on the Ξ°.

The origin of this rule is unknown, for if the decays are via some X coupled to ($\bar{p}n$), there is no apparent reason why the $\Delta T = 1/2$ and $\Delta T = 3/2$ amplitudes should not be of the same general order of magnitude. We are left also with the mystery of the origin of the small coefficients in X.

One speculation made by the authors (unpublished data) is that one particular diagram is very much more important than the others and about ten times as big as one would estimate. Let us illustrate this idea for the example that X is simply 0.1 ($\bar{p}\Lambda$). Then in the coupling $\bar{p}n \leftrightarrow \bar{p}\Lambda$ the $\bar{p}$ can be eliminated, giving a direct amplitude for transformation of $n \leftrightarrow \Lambda$. The

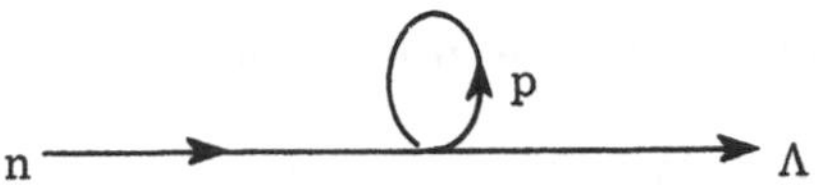

FIG. 14-1

mechanism of such an elimination is the diagram in Fig. 14-1, with the proton in a closed loop. We imagine this diagram with the loop undisturbed is much larger than one would expect, and larger than all other diagrams. The large size compensates the small coefficient 0.1 in X, and makes nonleptic decays appear at a normal rate. Further since the dominant coupling is now effectively $n \rightarrow \Lambda$, a transformation for which ΔT is restricted to $\pm 1/2$, this restriction appears as rule 5. The $\Delta T = 3/2$ can come only from the more complicated usual diagrams, for which the small coefficient 0.1 is not compensated.

But we explain the two mysteries (4 and 5) by two ad hoc assumptions (that the X coefficient is small and that a certain diagram is large), so it is not clear that we are getting anywhere. It is true that all the Λ and Σ decay details come out quite closely if we take the coupling to be equally $n \leftrightarrow Y$ and $p \leftrightarrow \Sigma^+$ using the global symmetry hypothesis and a perturbation calculation, but the perturbation approximation cannot be justified.

Summary. At this point it would be well to summarize the salient features of the problem we face.

According to the principles of quantum field theory there exist in nature only particles having mass and spin (even or odd half-integral), which have a relation among each other called coupling. These particles fall into two groups: the weakly and the strongly interacting particles. The weakly interacting particles include: the photon, the graviton (which everyone ignores), and the leptons (e, μ, ν). The strongly interacting particles are the mesons and the baryons (see Tables 9-1 and 9-2).

You know the particles and the couplings and so you know everything. Physics in a nutshell! Couplings other than gravitation (which conserves everything or nothing depending on how you look at it) have the properties listed in Table 14-1.

TABLE 14-1

Coupling	Relative strength	Conservation laws satisfied		
		i spin	Strangeness	Parity
Fermi	10^{-10}	No	No	No
Electrodynamics	10^{-2}	No	Yes	Yes
Strong	10^{1}	Yes	Yes	Yes

The conservation laws satisfied by all couplings are of two types:

Geometrical laws:

Angular momentum (rotation)

Energy, momentum (translation)

Parity × charge conjugation

Time reversal

Number laws:

Number of leptons

Electric charge

Number of baryons (nucleonic charge conservation)

We see that there are 31 particles; hence, according to standard field theory, we need 31 fields to describe these particles! However by being clever we can try to reduce the number of fields we need, for some of the particles may be compounds. What is the minimum number needed? First we need a baryon. Also we need something with i spin 1/2 (two states), and we need the quality of strangeness. So, for example, we could get by with three baryons: n, p, Λ. But this tells us nothing about leptons. We need then ν, e, γ and probably the graviton. No one is quite sure what to do with the μ. If we also include it we then have to explain at least eight fields and four couplings.

This is as far as one has been able to proceed, so far. It is clear that very considerable progress is made by noticing symmetries, but that no calculational work has otherwise yielded much information. We are sorely in need of reliable methods for a quantitative analysis of these problems.

But even more profound and exciting is the problem that is tacitly implied throughout these lectures but never clearly stated. What is the significance or the pattern behind all these interrelated symmetries, partial symmetries, and asymmetries?

15 Fundamental Laws of Electromagnetic and β-Decay Coupling

We shall now see how to make quantitative calculations for those processes that we can calculate! I shall just tell you the results and give heuristic arguments for their correctness. I don't feel that it is necessary to start from field theory since it is, in fact, not internally consistent. Anyway, I want to leave room for new ideas.

You may have great pedagogical difficulty in learning the physics presented in this way. It would be easier to learn it more historically, by moving from the Schrödinger equation to the Dirac equation, and from the quantization of harmonic oscillators to the creation and annihilation operators, and finally to the resulting amplitude for various processes. Instead, we shall just give rules for finding the resulting amplitude—because the rules are so much simpler than the steps leading to them. Furthermore, the things we would otherwise start with (e.g., the Schrödinger equation) are approximations to the end result, useful only under certain conditions. What is ultimately desired for a true physical understanding is how the Schrödinger equation is a consequence of the more fundamental laws. It is true that for historical or pedagogical understanding it would be better to start from Schrödinger and go the other way—although of course no real deduction can be made that way; new things like Dirac matrices, etc., must be added from time to time. But it is a long, hard climb to the frontier of physics that way. Let's make an effort to try an experiment in learning. Let us see if we can put you right down on the frontier so you can do two things. First, you can look forward into the unknown and see the problems and the progress being made and perhaps help to solve some of them. Second, you can look back and try to see how the various things you have learned, from Newton's laws to Maxwell's equations and to Schrödinger's equations, are all consequences of what you are learning now. The latter will not be obvious, and will make it hard to accept the apparently ad hoc rules we discuss now. But really, that is how nature does it; she "understands" the Schrödinger equation as an approximate equation describing large numbers of interactions among many particles moving slowly. The fundamental elements are the key interactions among small numbers of particles moving at arbitrary speed. To these we address our attention.

The only interactions for which a reasonably satisfactory quantitative quantum mechanical description can be given today are the electromagnetic coupling and the Fermi or β-decay coupling.

We consider processes that involve a small number of particles which may interact, create other particles, decay, etc. For each such process we have an amplitude; the square of the amplitude gives the probability for the process to occur. We start with the case where there are no virtual particles. This is easier than the processes involving virtual particles, which we shall treat later. Also we begin with spin 0 (scalar) particles, in order not to confuse things by introducing spin and relativity at the same time.

The wave function of the scalar particle has only one component. Under the transformation $x_\mu \rightarrow x'_\mu$ (rotation or Lorentz transformation),

$$\varphi(x,y,z,t) \rightarrow \varphi(x',y',z',t')$$

How about space reflection?

$$\mathbf{x} \rightarrow -\mathbf{x}$$

$$t \rightarrow t$$

If $\varphi(\mathbf{x},t) = \varphi(-\mathbf{x},t)$, we speak of a "scalar" particle; $\varphi(\mathbf{x},t) = -\varphi(-\mathbf{x},t)$ is a "pseudoscalar." Of course φ might not obey either of the above equations.

We shall assume that a free particle is represented by the plane wave

$$u \exp(-ip \cdot x)$$

where $p \cdot x \equiv p_\mu x_\mu = Et - \mathbf{p} \cdot \mathbf{x}$. This is DeBroglie's assumption; u does not change under coordinate transformations.

Next we seek an expression for the probability. This has to be the fourth component of a four-vector (S_μ), since the probability of finding the particle somewhere:

$$\int S_4 \, dx \, dy \, dz$$

must be invariant. Here S_4 = the probability of finding the particle per cm^3 (also the probability of passing from past to future), and $\mathbf{S}$ = the probability of passing through a surface normal to $\mathbf{S}$ per cm^2 sec.

$\varphi^* \varphi$ is a scalar, and hence not a satisfactory S_4. The space integral of S_4 is really a surface integral in four dimensions (see Fig. 15-1). The generalization is evidently

$$\text{Probability to pass through surface} = \int S_\mu N_\mu \, d^3 \text{ surface}$$

where N_μ is the unit normal to the surface, $N_\mu N_\mu = -1$.

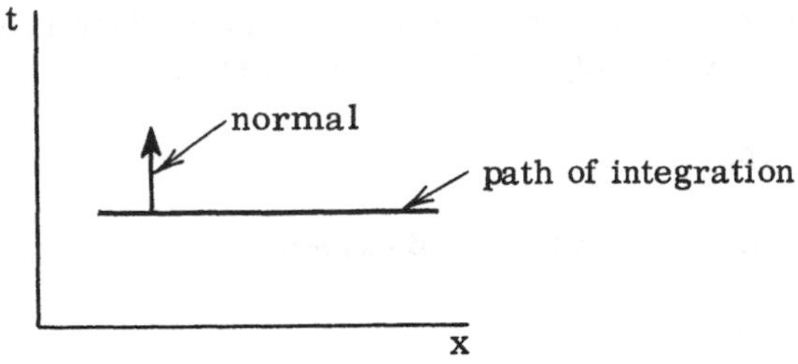

FIG. 15-1

Suppose the particle is located in some finite region of space (so that u is not a plane wave). Can somebody else in another Lorentz frame perform the probability integral in his frame and get the same result we do? Recalling that in the space-time diagrams the "moving" system is rotated, we draw the schematic picture of Fig. 15-2.

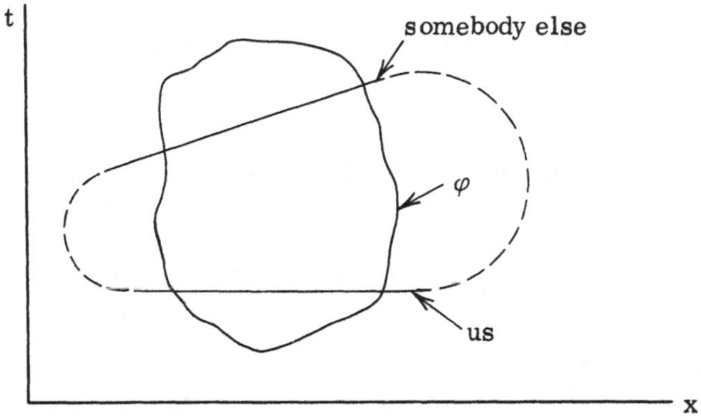

FIG. 15-2

Since the particle is localized as shown, S_μ is zero at some distance, so that we can close the path of integration indicated by the dotted lines in the figure.

Then both observers get the same answer if

$$\int S_\mu N_\mu \, d^3 \text{ surface} = 0$$

Using Gauss' theorem, this is assured if

$$\partial S_\mu / \partial x_\mu = 0$$

which expresses the conservation of probability.

We have seen that for a plane wave, $\bar{u}u$ is unsatisfactory for a probability density. Since p_μ is the only available four-vector,

$$S_\mu = 2p_\mu \bar{u}u$$

(The factor 2 is just convention.) This gives

$$S_4 = 2E\bar{u}u$$

Does this make sense? Note that the density in a moving system appears greater by just the same factor the E has. Thus keep $\bar{u}u = 1$ in every system; the relativistic normalization is then 2E per cm^3. This is a crazy normalization but very useful in practice. We shall always normalize in this way.

How about a more general expression for S_μ? For the plane wave

$$S_\mu = 2\bar{\varphi}\ i \frac{\partial}{\partial x_\mu} \varphi$$

In the general case we must take the more symmetrical forms

$$S_\mu = \bar{\varphi}\ i \frac{\partial}{\partial x_\mu} \varphi - i \frac{\partial \bar{\varphi}}{\partial x_\mu} \varphi$$

Now that we have everything defined, we shall see how to calculate. Recall the famous formula for the transition probability per second:

$$\text{Prob./sec} = 2\pi\, |M_{fi}|^2 \frac{\text{(density of final states)}}{\text{(unit energy interval)}}$$

evaluated for $E_f = E_i$.

This form is inconvenient for our purposes. I shall rewrite it so that you cannot recognize it. First, in order to use our normalization we must divide by 2E for each particle entering in the process. Since we shall always be working in a continuum of states we have

$$\text{Prob./sec} = 2\pi \frac{|M|}{\prod_{\text{in}} (2E) \prod_{\text{out}} (2E)} \frac{d^3 p_1}{(2\pi)^3} \frac{d^3 p_2}{(2\pi)^3} \cdots \frac{d^3 p_{N-1}}{(2\pi)^3} \frac{1}{dE}$$

Note that the expression for the density of states lacks $d^3 p_N/(2\pi)^3$ because of momentum conservation. To make the formula symmetrical among all the particles, we add the factor

$$[d^3 p_N/(2\pi)^3]\, (2\pi)^3\, \delta^{(3)} \left(\sum_{\text{in}} p - \sum_{\text{out}} p\right)$$

Also we replace $1/dE$ by $\delta\left(\sum_{\text{in}} E - \sum_{\text{out}} E\right)$. In fact we could have started with the formula in this form: Between any two states,

$$\text{Prob./sec.} = 2\pi\, |M_{fi}|^2\, \delta(E_f - E_i)$$

Further recall that

$$\delta(ax) = (1/a)\,\delta(x)$$

$$\delta(f(x)) = [1/|f'(x_0)|]\,\delta(x - x_0) \qquad \text{if } f(x_0) = 0$$

This allows us to remove the lack of symmetry between p and E in our formula, since

$$\int \delta(p_4^2 - \mathbf{p}^2 - m^2)dp_4 = 1/2p_4\Big|_{p_4 = (\mathbf{p}^2 + m^2)^{1/2}} \; 1/2E$$

Thus

$$d^3\mathbf{p}_i/2E_i(2\pi)^3 \rightarrow [d^3\mathbf{p}_i/(2\pi)^3]\; d\mathbf{p}_{4i}\; [\delta(p_i^2 - m^2)]$$
$$= d^4p_i/(2\pi)^4\,[2\pi\,\delta(p_i^2 - m^2)]$$

where $p_i = (E_i, \mathbf{p}_i)$.

Collecting all these factors we finally have

$$\text{Prob./sec} = [|M|^2/\text{product} \prod_{\text{in}}(2E)]\,(2\pi)^4\,\delta^{(4)}(\sum_{\text{in}} p - \sum_{\text{out}} p)$$

$$\times \prod_{\text{out}}(2\pi)\,\delta(p_i^2 - m_i^2)[d^4p_i/(2\pi)^4]$$

The factor $\delta^{(4)}(\sum_{\text{in}} p - \sum_{\text{out}} p)$ corresponds to over-all conservation of energy and momentum.

We shall always be going back and forth from configuration space to momentum space. Our convention is as follows:

$$\varphi(x) = \int u(p)\,\exp(-ip\cdot x)[d^4p/(2\pi)^4]$$

$$\varphi(p) = \int \exp(ip\cdot x)\,\varphi(x)\,d^4x$$

With this convention dp is always accompanied by a factor $1/2\pi$ and the δ function by 2π.

There is one more important remark to be made. The great utility of this method is that M turns out to be invariant. So we can choose the system in which we evaluate M at our convenience.

As an example, consider the disintegration of a particle. The probability of decay when it is moving is less by a factor M/E, which is of course the relativistic time dilation. I suggest that you try the following example, which we shall do in detail later. Consider the decay of a kaon into two π's:

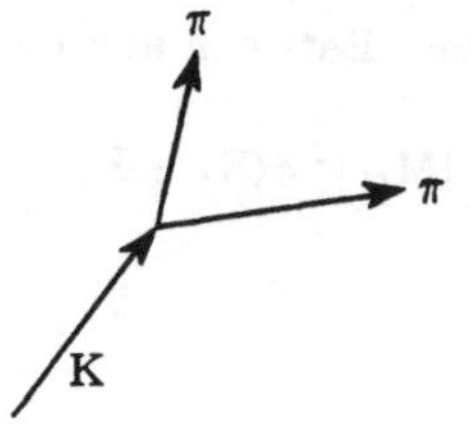

(Forget about charge just now.) The K and π have spin 0 and amplitude u_k, u_π. Assume that the amplitude for the process is

$$M = (4\pi)^{1/2} f\, M_K u_K u_\pi^* u_\pi^*$$

The factor $(4\pi)^{1/2}$ is conventional (rationalized units), f measures the strength of the interaction, and M_k is inserted to make f dimensionless. The u's are just to remind us of the process (the u* 's are the particles created and the u's the particles destroyed) but they are set equal to unity in the calculations. Find the value of f that gives the experimental lifetime.

16 Density of Final States

Generally we shall need to consider only two cases for the incoming state: (1) decay of a particle (prob. trans./sec = 1/mean life), and (2) two particles in collision (prob. trans./sec = σv, where σ is the cross section and v is the relative velocity).

We write the transition probability per second in the form

$$\text{Prob. trans./sec} = 2\pi\,(M^2/\prod_{\text{in}} 2E \prod_{\text{out}} 2E)\,D$$

where

$$D = (1/2\pi) \prod_{\text{out}} 2E(2\pi)^4 \delta^4(\sum_{\text{in}} p - \sum_{\text{out}} p) \prod_{\text{out}} 2\pi\delta(p^2 - m^2)[d^4p/(2\pi)^4]$$

is the density of state per unit range. There are other useful expressions for D:

(1) two outgoing particles:

$$D = \frac{E_1 E_2}{(2\pi)^3} \frac{p_1^3\, d\Omega_1}{E_T p_1^2 - E_1(\mathbf{p}_T \cdot \mathbf{p}_1)} \qquad \begin{aligned} E_T &= E_1 + E_2 \\ \mathbf{p}_T &= \mathbf{p}_1 + \mathbf{p}_2 \\ E_1 &= (p_1^2 + m_1^2)^{1/2} \end{aligned}$$

when $m_2 \to \infty$, $D = [E_1 p_1/(2\pi)^3]\, d\Omega_1$. In the c.m. system,

$$D = [1/(2\pi)^3]\,[E_1E_2/(E_1 + E_2)]\, p_1\, d\Omega_1$$

(2) three outgoing particles:

$$D = \frac{E_2 E_3}{(2\pi)^6} \frac{p_2^3 p_1^2\, dp_1\, d\Omega_1\, d\Omega_2}{p_2^2(E_T - E_1) - E_2\mathbf{p}_2 \cdot (\mathbf{p}_T - \mathbf{p}_1)} \qquad \begin{aligned} E_T &= E_1 + E_2 + E_3 \\ \mathbf{p}_T &= \mathbf{p}_1 + \mathbf{p}_2 + \mathbf{p}_3 \end{aligned}$$

when $m_3 \to \infty$.

$$D = [1/(2\pi)^6]\, E_2 p_2 p_1^2\, dp_1\, d\Omega_1\, d\Omega_2$$

Consider now the kaon decay into two π's, which I suggested in the last lecture. I have assumed a direct coupling $K \rightarrow \pi, \pi$ just to illustrate the techniques we shall be using; we do not believe that this is actually a fundamental coupling in nature. A word about units: We have chosen $\hbar = c = 1$. Then

$$m = \text{mass} = \text{energy} = 1/\text{length} = 1/\text{time}$$

$$= m \qquad = mc^2 \qquad = 1/(\hbar/mc) = 1/(\hbar/mc^2)$$

For the electron,

$$m_e = 9.1 \times 10^{-28}\ \text{g} = 0.511\ \text{Mev} = 1/3.86\times 10^{-11}\ \text{cm}$$

$$= 1/1.288\times 10^{-21}\ \text{sec}$$

These numbers should be memorized. For the proton all quantities are multiplied by 1836, and similarly, for other particles of mass m by m/m_e.

At the end of a calculation it will always be clear which units m represents. We can still check dimensions, e.g., a lifetime must be proportional to m. Keeping track of $\hbar$ and c is a complete waste of time!

The first time we shall do the calculation the hard way. We have

$$M = (4\pi)^{1/2} f\, M_K\, u^*_{\pi_1} u^*_{\pi_2} u_K \qquad m_1 = m_2 = m_\pi$$

$$\text{Prob./sec} = 1/\tau = (4\pi f^2 M_K^2/2M_K)(2\pi)^4 \delta^4 (p_K - p_1 - p_2)\, 2\pi\delta(p_1^2 - m_\pi^2)$$

$$\times [d^4p_1/(2\pi)^4]\, (2\pi)\, \delta\, (p_2^2 - m_\pi^2)[d^4p_2/(2\pi)^4]$$

Setting $p_2 = p_K - p_1$ we can cross out $(2\pi)^4 \delta^4(p_K - p_1 - p_2)$ and $d^4p_2/(2\pi)^4$. Let $p_1 = (E, \mathbf{p}_-)$, $p_K = (M_K, 0)$ (in the rest frame of the K). Then

$$(p_K - p_1)^2 = p_K^2 - 2p_K p_1 + p_1^2 = M_K^2 - 2M_K E + m_\pi^2$$

and

$$\text{Decay rate} = 1/\tau = (2\pi) f^2 M_K\, (2\pi)\, \delta\, (E^2 - p^2 - m_\pi^2) 2\pi\delta(M_K^2 - 2M_K E)$$

$$\times\ [p^2\, dp\, d\Omega\, dE/(2\pi)^4]$$

Furthermore,

$$\int 2\pi\, \delta(M_K^2 - 2M_K E)\, dE = 2\pi/2M_K \qquad E = M_K/2$$

$$\int 2\pi\, \delta[(M_K^2/4) - p^2 - m_\pi^2]\, p^2\, dp = (2\pi/2)p \qquad p = [(M_K^2/4) - m_\pi^2]^{1/2}$$

$$d\Omega = 4\pi$$

so that

$$1/\tau = 2\pi\, f^2 M_K\, \pi[(M_K^2/4) - m_\pi^2]^{1/2} (\pi/M_K)[4\pi/(2\pi)^4]$$

$$= (f^2/4)\, M_K [1 - (2m_\pi/M_K)^2]^{1/2}$$

The whole purpose of physics is to find a number, with decimal points, etc.! Otherwise you haven't done anything. Find

$$1/\tau = f^2[(966 \times 0.84)/4 \times (1288 \times 10^{-21}) \text{ sec}]$$

Experimentally the lifetime of the K = 0.99×10^{-10} sec. Only 78 per cent of the decays go into 2π. Therefore,

$$1/\tau_{\text{exp}} = 0.78/0.99 \times 10^{-10} \text{ sec}$$

and

$$f = 2.38 \times 10^{-7}$$

This is a very small dimensionless constant so we are confronted with a weak coupling. I want to impress on you that this is just an example that we have made up; we do not believe that the true mechanism of this decay is this fundamental coupling, but rather that the decay occurs as an indirect effect of some other couplings. However, let's go along with it.

Problem 16-1: The K also decays into 3 π's. I am going to assume a coupling

$$(4\pi)^{1/2}\, g\, u_{\pi_1} u_{\pi_2} u_{\pi_3} u_K$$

Obtain the spectral distribution, check it with the experimental results, and determine g.

Consider now a slightly more difficult problem, namely $\pi - K$ scattering. Forget the charge of the K and the π, a possible way in which this could occur is illustrated in Fig. 16-1. This is an indirect process involving a virtual π. I shall now give you the rules for finding the amplitude for such a process (later on we shall make them more plausible).

Follow along the particle lines and write (from right to left):

(1) for each vertex an amplitude $(4\pi)^{1/2} f M_K$

(2) for the propagation of the π between two vertices an amplitude

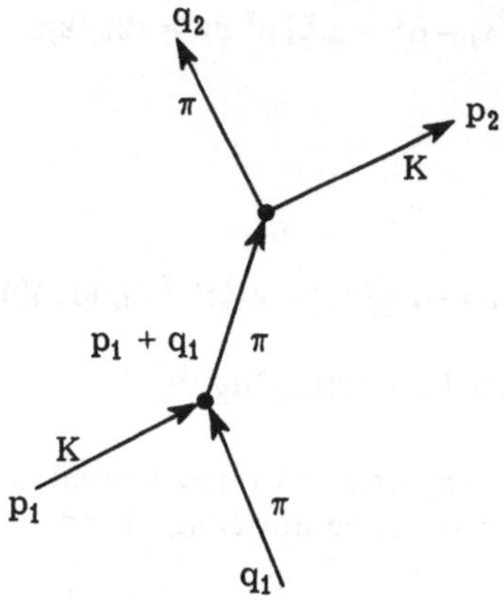

$K + \pi \rightarrow \pi \rightarrow K + \pi$

FIG. 16-1

$1/(p^2 - m_\pi^2)$, where p is the 4-momentum and m_π is the mass of the π. This is the equation of motion:

(3) energy momentum must be conserved at each vertex

The product of these amplitudes is then the amplitude M for the process.

For Fig. 16-1 we get

$$(4\pi)^{1/2} fM_K\{1/[(p_1 + q_1)^2 - m_\pi^2]\} (4\pi)^{1/2} fM_K$$

However there is another way in which the same transition can occur, shown in Fig. 16-2, which is topologically different from the first diagram (the vertices cannot be arranged in space time to get the first diagram). The amplitude for this process is

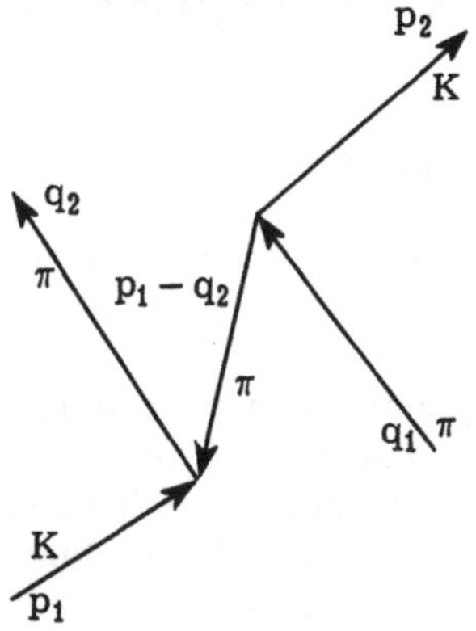

FIG. 16-2

$$(4\pi)^{1/2} fM_K \{1/(p_1 - q_1)^2 - m_\pi{}^2]\} (4\pi)^{1/2} fM_K$$

and the total amplitude M for the transition is obtained by adding these two:

$$M = 4\pi f^2 M_K{}^2 (\{1/[(p_1 + q_1)^2 - m_\pi^2]\} + \{1/[(p_1 - q_2)^2 - m_\pi^2]\})$$

17 The Propagator for Scalar Particles

I shall try to make the rule for the propagator less artificial by connecting it with something you already know. Consider the $\pi - K$ scattering example (Fig. 17-1).

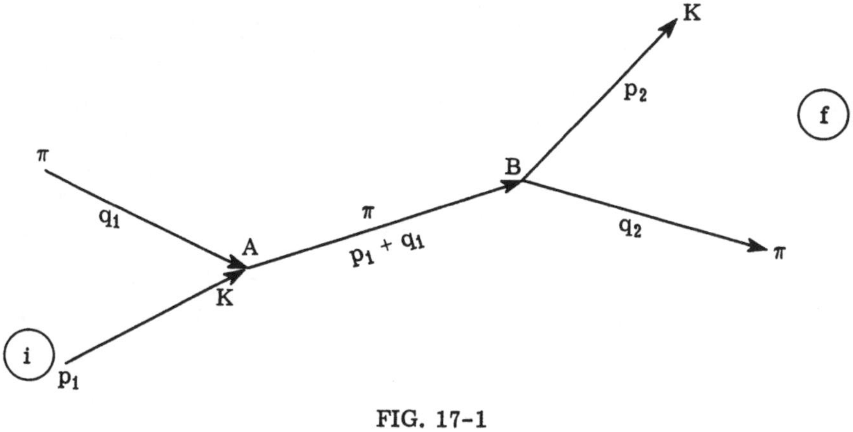

FIG. 17-1

We have been saying that the amplitude has the form

$$M = B_{fp}[1/(p^2 - m^2)]\,A_{pi}$$

Now consider the lowest-order term in ordinary perturbation theory:

$$\sum_n H_{fn}\,[1/(E_i - E_n)]\,H_{ni}$$

The sum is to be taken over the intermediate states n. The contribution of the diagram in Fig. 17-1 is

$$B_{fp}[1/(E_i - E_p)]\,A_{pi}$$

where

$$E_p = (\mathbf{p}^2 + m^2)^{1/2} \qquad \mathbf{p} = \mathbf{p}_1 + \mathbf{q}_1$$

But we must remember that in ordinary perturbation theory the process with opposite time order is considered separately as pair production followed by annihilation of the positron with the incoming electron. The energy in the intermediate state is thus $2E_i + E_p$ (Fig. 17-2). Recalling our pre-

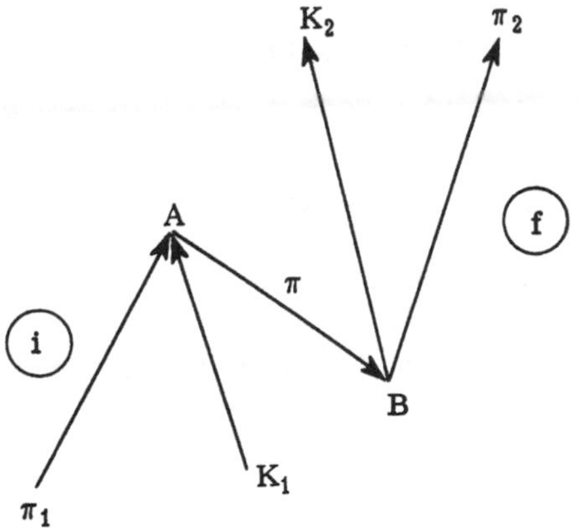

FIG. 17-2

vious rule for entry and exit states we see that f(i) is the entry (exit) state. So we have

$$A_{pi}\,\{1/[E_i - (2E_i + E_p)]\}\,B_{fp} = -B_{fp}[1/(E_i + E_p)]\,A_{pi}$$

The sum of the two relevant matrix elements is

$$2E_p\,B_{fp}[1/(E_i^2 - E_p^2)]\,A_{pi}$$

Now

$$\begin{aligned} p^2 - m^2 &= (p_1 + q_1)^2 - m^2 \\ &= (\varepsilon_1 + K_1)^2 - \mathbf{p}^2 - m^2 \\ &= E_i^2 - E_p^2 \end{aligned}$$

and the factor $2E_p$ is just the normalization constant. So the backward-in-time idea simplifies the result. Each of the above terms (considered separately) is not invariant. By combining them we obtain an obviously invariant expression. This demonstration is not intended to be a proof of our propagator rule, but does show the physical notions involved.

There is another way to get our propagator rule. For a free particle a special solution is

$$u \exp(-ip \cdot x) \qquad p^2 = m^2$$

We can form any free-particle solution by the sum

$$\varphi(\mathbf{x},t) = \Sigma_{\mathbf{p}}\, u(\mathbf{p}) \exp(-ip \cdot x)$$

Note that

$$\Sigma_{\mathbf{p}}(p^2 - m^2)u(\mathbf{p}) \exp(-ip \cdot x) = 0$$

But this is the same as

$$\sum\{[i(\partial/\partial t)]^2 - (-i\boldsymbol{\nabla})^2 - m^2\}u(\mathbf{p}) \exp(-ip \cdot x)$$

$$[(i\nabla_\mu)^2 - m^2]\varphi = 0$$

or

$$(\Box^2 - m^2)\varphi = 0$$

where

$$\Box^2 \equiv \boldsymbol{\nabla}^2 - (\partial^2/\partial t^2)$$

Suppose there is a source $S(\mathbf{x},t)$ for the particles. We then postulate that

$$(\Box^2 - m^2)\,\varphi(\mathbf{x},t) = S(\mathbf{x},t)$$

We solve this equation by introducing the Fourier transforms

$$S(\mathbf{x},t) = \int \exp(-ip \cdot x)\, S(p)[d^4p/(2\pi)^4]$$

$$\varphi(\mathbf{x},t) = \int \exp(-ip \cdot x)\,\varphi(p)\,[d^4p/(2\pi)^4]$$

(Note that $p_4 \neq (p^2 + m^2)^{1/2}$ since we are no longer dealing with a free particle.)

The transformed equation is

$$(p^2 - m^2)\varphi(p) = S(p)$$

So if we know the source, then

$$\varphi(p) = [1/(p^2 - m^2)]\, S(p)$$

showing the origin of the propagator. Solving for $\varphi(\mathbf{x},t)$,

$$\varphi(\mathbf{x},t) = \int \exp(-ip\cdot x)\,[1/(p^2 - m^2)] \int \exp(-ip\cdot x')S(x')\, d^4x' \times [d^4p/(2\pi)^4]$$

or

$$\varphi(x) = \int D_+(x - x')S(x')\, d^4x'$$

where

$$D_+(x - x') = \int [\exp(-ip)(x - x')/(p^2 - m^2)]\,[d^4p/(2\pi)^4]$$

is the propagator in space and time.

We notice that in order to give a meaning to D_+ it is necessary to define the pole in the integrand. To do this we add an infinitesimal negative imaginary part to the mass (an invariant) and integrate first over $d\omega = dp_4$:

$$\int \frac{\exp[i\mathbf{p}\cdot(\mathbf{x} - \mathbf{x}')]\,\exp[-i\omega(t - t')]}{\omega^2 - p^2 - m^2 + i\epsilon}\,\frac{d\omega}{2\pi}\,\frac{d^3p}{(2\pi)^3}$$

The limit $\epsilon \to 0$ is understood to be taken afterward.

This prescription displaces the poles from $\omega = \pm E_{\mathbf{p}} = \pm(\mathbf{p}^2 + m^2)^{1/2}$ to $\omega = \pm E_{\mathbf{p}} \mp i\epsilon$ and is equivalent to the contour in Fig. 17-3. For $t > t'$ we

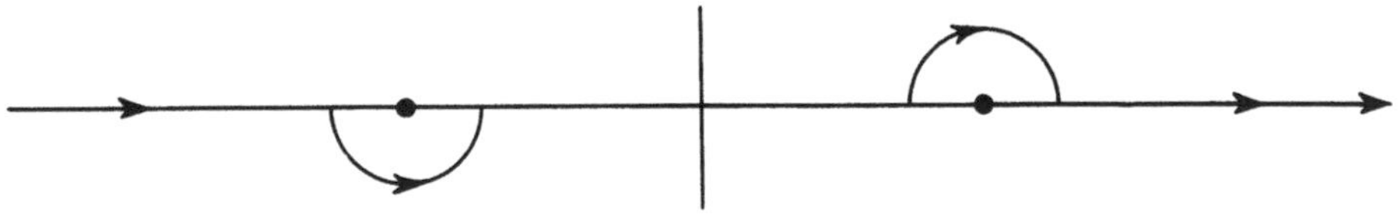

FIG. 17-3

complete the contour in the lower half-plane. Therefore,

$$D_+(t > t') = -2\pi i\,\mathrm{Res}(E_{\mathbf{p}}) = i\,[\exp(iE_{\mathbf{p}})(t - t')/E_{\mathbf{p}}]\,\exp[i\mathbf{p}\cdot(\mathbf{x} - \mathbf{x}')] \times [d^3\mathbf{p}/(2\pi)^3]$$

We note that for $t > t'$ only positive energies contribute. For $t < t'$ we must close the contour in the upper half-plane. We get

$$D_+(t < t') = i \int [\exp(iE_{\mathbf{p}}(t - t')/E_{\mathbf{p}}] \exp(i\mathbf{p} \cdot (\mathbf{x} - \mathbf{x}')) [d^3p/(2\pi)^3]$$

showing that only negative energies contribute for $t < t'$. In this way we see how the $m \rightarrow m - i\epsilon$ rule summarizes our prescriptions for going backward and forward in time.

Finally, the correct formula for the propagator of a spin-0 particle is

$$1/(p^2 - m^2 + i\epsilon)$$

18 The Propagator in Configuration Space

We have seen in the $\pi - K$ scattering example that in ordinary second-order perturbation theory the separate contributions to the amplitude were not relativistically invariant, but that their sum was invariant. Two Lorentz frames A and B are related by a rotation in space time. For example, the time order of the two successive interactions would be different in A and B in the situation in Fig. 18-1. This is true when the second vertex does not

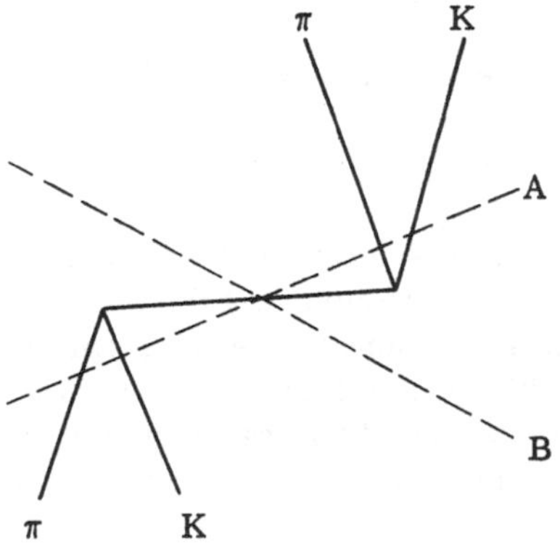

FIG. 18-1

lie within the light cone of the first (otherwise the time order could not be reversed by a Lorentz transformation).

One might think there is 0 amplitude if the two events are separated in space but not time. This is not the case. All positions contribute.

To see that this makes sense we examine some of the properties of the propagator in configuration space:

$$D_+(x) = \int [\exp(-ip \cdot x)/(p^2 - m^2 + i\epsilon)] [d^4p/(2\pi)^4]$$

How does $D_+(x)$ behave with respect to the light cone? In configuration space the propagator is much more complicated than in momentum space. Explicitly,

$$D_+(x) = -(1/4\pi)\delta(s^2) + (m/8\pi s)H_1^{(2)}(ms)$$

where

$$s = (t^2 - x^2)^{1/2} \qquad \text{for } t^2 > x^2$$

$$= -i(x^2 - t^2)^{1/2} \qquad \text{for } t^2 < x^2$$

$H^{(2)}$ is a Hankel function of the second kind.[9] For large s,

$$D_+ \approx (2/\pi s)^{1/2} \exp(-ims)$$

Note that for low velocities using $s \cong t - x^2/2t$,

$$D_+ \propto \exp(-imt) \exp[i(mx^2/2t)]/t^{3/2} = \exp(-imt)\psi_S$$

where ψ_S solves the Schrödinger equation. Outside the light cone for $\mathbf{x}^2 > t^2$, D_+ dies off exponentially:

$$D_+ \propto \exp(-m\sigma) \qquad \sigma = (x^2 - t^2)^{1/2}$$

$$\rightarrow \exp(-mr) \qquad \text{for } t^2 \ll x^2$$

As a physical illustration suppose that we measure the position of an electron, with a shutter, for example. At the same time, but at a different position, we make a measurement to see if we can find the electron there (Fig. 18-2). The probability is *not* zero, because in the act of measurement a pair could be created, the positron then annihilating the original electron. Pauli invented this thought experiment after he had thought the idea was wrong.

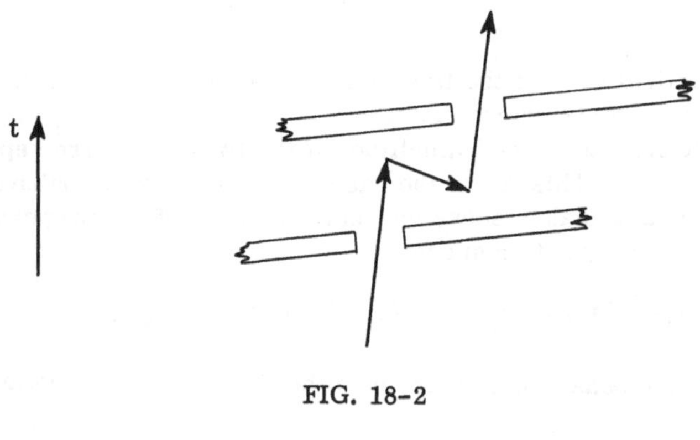

FIG. 18-2

Now consider a particle of high velocity; see Fig. 18-3.

Consider the behavior of the amplitude (Fig. 18-4) as we go across the light cone along AP. Now, does the wavelength at the point A (for example) correspond to the classical velocity x/t? Let us examine the phase of

$$\exp[-im(t^2 - x^2)^{1/2}]$$

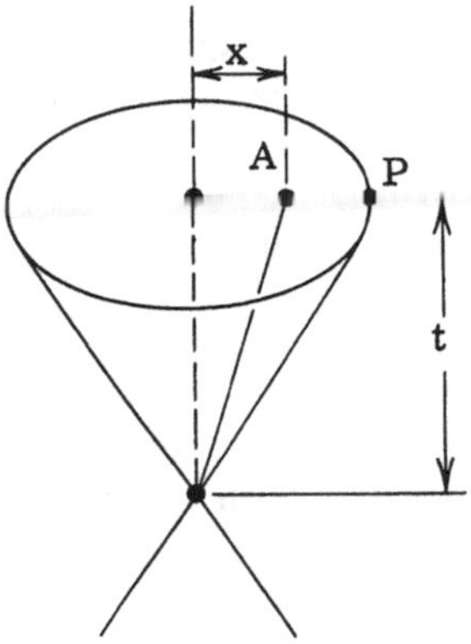

FIG. 18-3

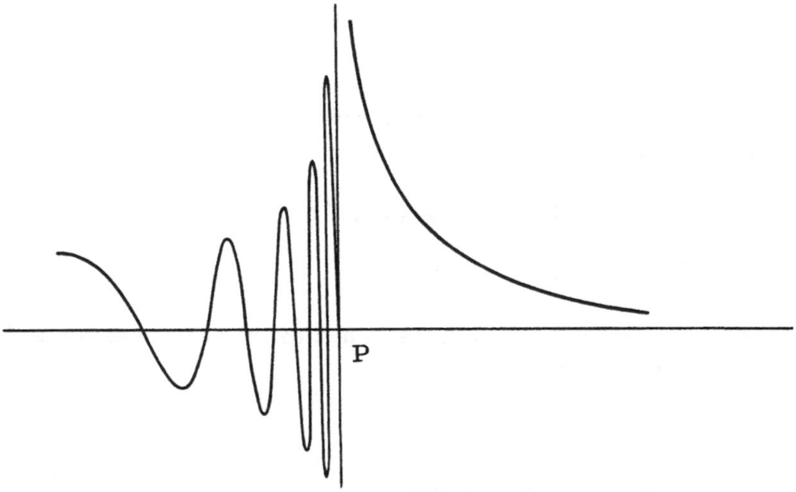

FIG. 18-4

When x changes by λ the phase must change by 2π:

$$m[t^2 - (x + \lambda)^2]^{1/2} - m(t^2 - x^2)^{1/2} = 2\pi$$

or

$$m\lambda\,(\partial/\partial x)(t^2 - x^2)^{1/2} = 2\pi$$

Therefore,

$$2\pi/\lambda = K = m(x/t)/[1-(x/t)^2]^{1/2} = mv_{\text{class}}/(1 - v_{\text{class}}^2)^{1/2}$$

We note that as $x \to t$ ($v \to c$) we approach a δ singularity on the light cone. A possible physical reason for this is that all momenta contribute, but for most momenta v is nearly c, so there is a great accumulation of amplitude near the light cone.

We have written the equation of motion for spin-0 boson fields as

$$(\Box^2 - m^2)\varphi = S$$

Now we must consider what S is; i.e., what can create the particles φ? We consider the $\pi - K$ example again with the coupling

$$(4\pi)^{1/2} f\,\varphi_\pi \varphi_\pi \varphi_K$$

Setting $f' = (4\pi)^{1/2} f$, the equations for φ_π and φ_K are

$$(\Box^2 - m^2)\varphi_\pi = 2f'\varphi_\pi\varphi_K$$

$$(\Box^2 - m^2)\varphi_K = f'\varphi_\pi^2$$

It is possible to obtain these results from a principle of least action. Consider the action

$$S = (1/2)\int [(\nabla_\mu \varphi_\pi)^2 - m_\pi^2 \varphi_\pi^2]\, d^4\tau$$

$$d^4\tau \equiv dx\, dy\, dz\, dt$$

Variation of φ_π, integration by parts (omitting the surface term), and setting $\delta S = 0$ gives the equation

$$-\nabla_\mu^2 \varphi_\pi - m^2\varphi_\pi = 0$$

This is the equation for the free pion.

For the $\pi - K$ example we add a similar term for the free K and an interaction term:

$$S = \int\{(1/2)[(\nabla_\mu\varphi_\pi)^2 - m_\pi^2\varphi_\pi^2] + (1/2)[(\nabla_\mu\varphi_K)^2 - m_K^2\,\varphi_K^2] + f'\varphi_\pi^2\varphi_K\}\, d\tau_4$$

Variation with respect to φ_π, φ_K yields the equations of motion given above. (Here we have tacitly assumed a real field describing neutral particles—the generalization is easy.)

More generally we have

$$S = \int \mathcal{L}\, d^4\tau$$

when the Lagrangian density $\mathcal{L}$ must be relativistically invariant. This requirement greatly restricts the number of admissible Lagrangian densities. Note the relation to the usual classical form,

$$S = \int L\, dt \qquad \text{with } L = \int \mathcal{L}\, d^3\mathbf{x}$$

We regard the action to be the more fundamental quantity. From it we can immediately read off the rules for the propagators, the coupling, and the equations of motion. But we still do not know the reason for the rules for the diagrams, or why we can get the propagators out of S.

19 Particles of Spin 1

In general we want to find amplitudes which transform under Lorentz transformations linearly:

$$u' = \mathcal{D}(L)\,u \qquad (L = \text{Lorentz transformation})$$

where

$$\mathcal{D}(L_1L_2) = \mathcal{D}(L_1)\,\mathcal{D}(L_2)$$

One solution is scalar. We can easily find another; 4-vectors transform linearly—so surely a 4-vector is a possible answer. We note the same for rotations—a 3-vector representing angular momentum 1 was permissible. Thus a particle can be represented by an amplitude which is a 4-vector. We expect it to have spin 1. There is just one complication, however: Under rotations the space components transform like a vector, but the time component transforms like a scalar. So it looks as if we are representing two particles. We can get around this difficulty by requiring that

$$p_\mu u_\mu = 0$$

Then in the rest frame of the particle ($\mathbf{p} = 0$)

$$mu_4 = 0 \qquad \text{or} \qquad u_4 = 0$$

Photons. The photon is the only known elementary particle of spin 1. It has 0 mass. If we know the laws of propagation of photons and their coupling to other particles, then we know all about electrodynamics. A very useful guide in formulating these laws is the requirement that in the classical limit the theory correspond to Maxwell's equations.

The amplitude for finding photons in quantum electrodynamics is taken to be the 4-vector potential $A_\mu(x,y,z,t)$, which in the absence of sources satisfies the equation

$$\Box^2 A_\mu = 0 \tag{19-1}$$

$$\nabla_\mu A_\mu = 0 \tag{19-2}$$

A free photon is represented by a plane wave

$$\varepsilon_\mu \exp(iKx)$$

where ε_μ is called the polarization vector. Substituting in Eq. (19-1) we find that $K^2 = 0$, or $m = 0$, and from Eq. (19-2) we find $K_\mu \varepsilon_\mu = 0$; the polarization is perpendicular to K_μ.

The theory must also be gauge invariant: If someone solves a problem with A_μ and somebody else with $A'_\mu = A_\mu + \nabla_\mu x$ where $\Box^2 x = 0$, both should get the same physical result. For plane waves this tells us something like this: Let $A_\mu = \varepsilon_\mu \exp(-iKx)$ represent a photon of momentum K and polarization ε and

$$x = i\alpha \exp(-iKx) \qquad \alpha = \text{const}$$

Then

$$A'_\mu = \varepsilon'_\mu \exp(-iKx) \qquad \varepsilon'_\mu = \varepsilon_\mu + \alpha K_\mu$$

Therefore, if two polarization vectors differ only by a multiple of the 4-momentum, they must represent the same photon. By a suitable gauge transformation we can always choose $\varepsilon_4 = 0$. Suppose $\varepsilon_4 \neq 0$. Let

$$\alpha = -\varepsilon_4/K_4$$

Then

$$\varepsilon'_4 = \varepsilon_4 - (\varepsilon_4/K_4)K_4 = 0$$

and

$$K_\mu \varepsilon'_\mu = \mathbf{K} \cdot \boldsymbol{\varepsilon}' = 0$$

A free photon is therefore represented by only two states of polarization. We can choose for these any two directions perpendicular to its momentum, or resolve them into right- and left-handed circular polarization (see Lecture 2). RHC (LHC) correspond to spin 1 along (opposite) the momentum of the photon. This can be seen easily as follows:

$$u_{\text{RHC}} = [1/(2)^{1/2}](\boldsymbol{\varepsilon}_x + i\boldsymbol{\varepsilon}_y)$$

where $\boldsymbol{\varepsilon}_x$ and $\boldsymbol{\varepsilon}_y$ are two unit vectors normal to the direction of propagation. Rotating θ degrees about the z axis,

$$u'_{RHC} = [1/(2)^{1/2}]\,(\boldsymbol{\varepsilon}'_x + i\boldsymbol{\varepsilon}'_y)$$

where

$$\boldsymbol{\varepsilon}'_x = \boldsymbol{\varepsilon}_x \cos\theta - \boldsymbol{\varepsilon}_y \sin\theta$$

$$\boldsymbol{\varepsilon}'_y = \boldsymbol{\varepsilon}_x \sin\theta - \boldsymbol{\varepsilon}_y \cos\theta$$

Substituting in u'_{RHC} we get $u'_{RHC} = \exp(i\theta)\, u_{RHC}$. In the same way,

$$u'_{LHC} = \exp(-i\theta)\, u_{LHC} \qquad u_{LHC} = [1/(2)^{1/2}]\,(\boldsymbol{\varepsilon}_x - i\,\boldsymbol{\varepsilon}_y)$$

Recall that the rotation matrix is $\exp(i\theta J_z)$. Therefore

$$J_z\, u_{RHC} = u_{RHC}$$

$$J_z\, u_{LHC} = -u_{LHC} \qquad \text{Q.E.D.}$$

We proceed now to find the laws of coupling and propagation of photons.

The Principle of Minimal Electromagnetic Coupling. There is a very interesting principle by means of which we can obtain the coupling of photons with a charged particle whenever the equation of motion of that particle is known. Take, for instance, the equation for a free scalar particle,

$$(i\nabla_\mu\, i\nabla_\mu - m^2)\varphi = 0$$

Then the rule is to change $i\nabla_\mu$ to $i\nabla_\mu - eA_\mu$. This gives an equation which contains the effects of the electromagnetic field:

$$[(i\nabla_\mu - eA_\mu)(i\nabla_\mu - eA_\mu) - m^2]\,\varphi = 0$$

It is important to note that this principle keeps the equations gauge-invariant. Let

$$\varphi = \exp(iex)\,\varphi'$$

Then φ' satisfies the same equation as φ with A_μ replaced by $A_\mu + \nabla_\mu x$. But φ and φ' differ only by a phase factor (which, however, may depend on space time), and consequently they represent the same physical state.

We can write the equation for φ in the form

$$(i\nabla_\mu i\nabla_\mu - m^2)\varphi = e[i\nabla_\mu(A_\mu\varphi) + A_\mu(i\nabla_\mu\varphi)] - e^2 A_\mu A_\mu \varphi$$

The right-hand side is the source of the scalar field. We can obtain the rules for the amplitude of the fundamental processes as follows: The amplitude for a particle with momentum p_1 $[\varphi_1 = \exp(-ip_1x)]$ to emit a photon of momentum q and polarization $\mathcal{E}[A_\mu = \mathcal{E}_\mu \exp(iqx)]$ and continue with momentum $p_2[\varphi_2 = \exp(ip_2x)]$ is proportional to

$$e \int \varphi_2^* [i\nabla_\mu (A_\mu \varphi_1) + A_\mu (i\nabla_\mu \varphi_1)] \, d^4x$$

$$= e \int \exp(ip_2x)\{i\nabla_\mu [\mathcal{E}_\mu \exp(iqx) \exp(-ip_1x)] + \mathcal{E}_\mu \exp(iqx)$$

$$\times \, i\nabla_\mu \exp(-ip_1x)\} \, d^4x$$

$$= e(p_1 - q + p_1) \cdot \mathcal{E} \int \exp[i(p_2 + q - p_1)x] \, d^4x$$

The last factor expresses the conservation of energy and momentum at the vertex: $p_2 + q = p_1$. If the photon is absorbed, replace q by $-q$.

In either case the amplitude is given by

$$\text{Amp.} = -i(4\pi)^{1/2} e(p_2 + p_1) \cdot \mathcal{E}$$

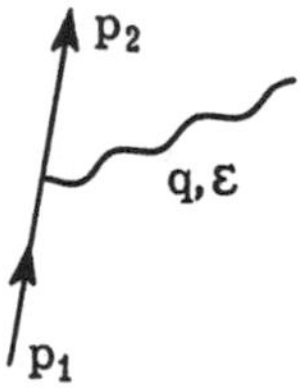

The factor $(4\pi)^{1/2}$ is introduced so that e is the unrationalized coupling constant $e^2 = 1/137$ in units where $\hbar = c = 1$. The factor $-i$ is essential in order to keep the correct phase relationship when indistinguishable processes which are of higher order in the coupling constant are included, but otherwise it can be left out.

The term quadratic in e gives the amplitude for the simultaneous emission (absorption) of two photons. The amplitude is proportional to

$$e^2 \int \varphi_2^* A_\mu A_\mu \varphi_1 \, d^4x$$

$$= e^2 \int \exp(ip_2x) [\mathcal{E}_\mu^a \exp(iq_ax) \, \mathcal{E}_\mu^b \exp(iq_bx) + \mathcal{E}_\mu^b \exp(iq_bx)$$

$$\times \, \mathcal{E}_\mu^a \exp(iq_ax)] \exp(-ip_1x) \, d^4x$$

$$= e^2 (\mathcal{E}_a \cdot \mathcal{E}_b + \mathcal{E}_b \cdot \mathcal{E}_a) \int \exp[i(p_2 + q_a + q_b - p_1)x] \, d^4x$$

The factor $\mathcal{E}_a \cdot \mathcal{E}_b$ appears twice, since either of the two A_μ's could have emitted photon a or photon b. Again, the last factor expresses the conservation of 4-momentum: $p_2 + q_a + q_b = p_1$. The amplitude is now

$$\text{Amp.} = -4\pi e^2 (\mathcal{E}_a \cdot \mathcal{E}_b + \mathcal{E}_b \cdot \mathcal{E}_a)$$

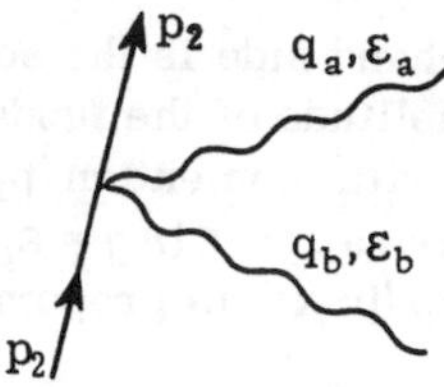

Let us emphasize again that the connection between the rules for amplitudes and the classical equations of motion is only heuristic. It is clearly impossible to "derive" quantum electrodynamics from Maxwell's equations; these can only serve as a guide.

Alternatively we could have started with the Lagrangian density of the free scalar field φ

$$\mathcal{L}_F = -(i\nabla_\mu \varphi)^* (i\nabla_\mu \varphi) + m^2 \varphi^* \varphi$$

Changing $i\nabla_\mu \rightarrow i\nabla_\mu - eA_\mu$ we get

$$\mathcal{L} = -(-i\nabla_\mu - eA_\mu)\varphi^* (i\nabla_\mu - eA_\mu)\varphi + m^2 \varphi^* \varphi$$

Expanding we can write

$$\mathcal{L} = \mathcal{L}_F + \mathcal{L}_c$$

where

$$\mathcal{L}_c = eA_\mu [(i\nabla_\mu \varphi)^* \varphi + \varphi^* (i\nabla_\mu \varphi) - e^2 A_\mu A_\mu \varphi^* \varphi$$

is the contribution due to the coupling between particles and photons. The rules for the amplitudes of the fundamental processes can also be read from $\mathcal{L}_c$.

The coefficient of e tells us for instance, that there is a process in which a particle with momentum $p_1 [\varphi = \exp(-ip_1x)]$ emits a real or virtual photon with momentum q and polarization $\mathcal{E} [A_\mu = \mathcal{E}_\mu \exp(iqx)]$ and goes on with momentum $p_2 [\varphi = \exp(-ip_2x)]$. Substituting in $\mathcal{L}_c$ we get

$$e \int \mathcal{E}_\mu \exp(iqx)[p_{2\mu} \exp(ip_2x) \exp(-ip_1x) + \exp(ip_2x)\, p_{1\mu} \exp(-ip_1x)] \times d^4x$$
$$= e[p_2 + p_1] \cdot \mathcal{E} \int \exp[+i(q + p_2 - p_1)x]\, d^4x$$

The last factor tells us that $p_2 + q = p_1$. The amplitude is then

$$\text{Amp.} = -i(4\pi)^{1/2} e(p_2 + p_1) \cdot \mathcal{E}$$

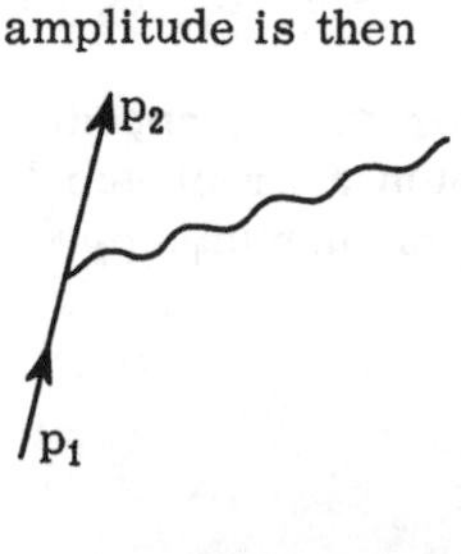

The inclusion of the factor $(4\pi)^{1/2}$ and $-i$ was discussed previously.

The coefficient of e^2 corresponds to the simultaneous emission of two photons: One of the A_μ's is

$$\varepsilon_\mu^a \exp(iq_a x) \qquad \text{or} \qquad \varepsilon_\mu^b \exp(iq_b x)$$

and correspondingly the other A_μ is

$$\varepsilon_\mu^b \exp(iq_b x) \qquad \text{or} \qquad \varepsilon_\mu^a \exp(iq_a x)$$

The amplitude for the process is therefore

$$\text{Amp.} = -4\pi e^2\, 2\varepsilon_a \cdot \varepsilon_b$$

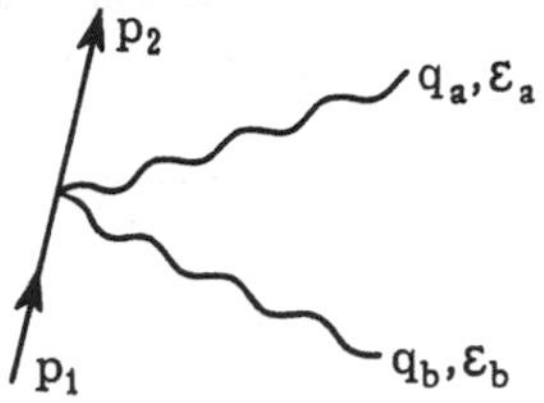

The Photon Propagator. The photon propagator can also be obtained from the equations of motion. This amplitude $A_\mu(x,y,z,t)$ for a photon satisfies Maxwell's equation

$$\nabla_\mu \nabla_\mu A_\nu = j_\nu \qquad j_\nu \text{ is the source of photons.}$$

Since $\nabla_\nu A_\nu = 0$, it follows that $\nabla_\nu j_\nu = 0$; we shall say more about this later. Following the procedure described in Lecture 17 let

$$A_\mu(x) = \int \varepsilon_\mu(k) \exp(-ikx)\,[d^4k/(2\pi)^4]$$

$$j_\mu(x) = \int j_\mu(k) \exp(-ikx)\,[d^4k/(2\pi)^4]$$

Substituting in the differential equation we get

$$-k^2 \varepsilon_\mu(k) = j_\mu(k)$$

Consequently the propagator for virtual photons is

$$-(i/k^2)\,\delta_{\mu\nu}$$

The factor $\delta_{\mu\nu}$ serves to remind us what kind of sources produce what polarization, the factor i is included because of the factor $-i$ in the coupling.

As an example, consider the $\pi - K$ scattering via photons (Fig. 19-1). (Forget the direct $K\pi\pi$ interaction which we imagined previously.)

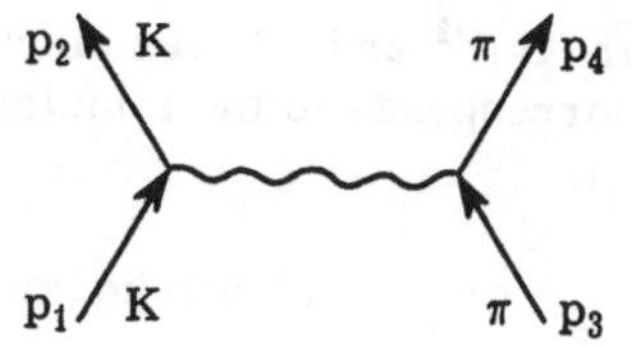

FIG. 19-1

The total amplitude M is made up of three factors:

(1) amplitude for the K meson of momentum p_1 to emit a virtual photon of momentum $p_1 - p_2$, polarization ε:

$$(4\pi)^{1/2}\, e\, (p_{1_\mu} + p_{2_\mu})\, \varepsilon_\mu$$

(2) amplitude for the photon propagation:

$$-[\delta_{\mu\nu}/(p_1 - p_2)^2]$$

(3) amplitude for the π meson with momentum p_3 to absorb the virtual photon:

$$(4\pi)^{1/2}\, e\, (p_{3_\mu} + p_{4_\mu})\, \varepsilon_\mu$$

On summing over all four directions of polarization of the virtual photons,

$$M = 4\pi e^2 \sum_{\text{pol}} (p_{1_\mu} + p_{2_\mu})\varepsilon_\mu (p_{3_\nu} + p_{4_\nu})\varepsilon_\nu [\delta_{\mu\nu}/(p_1 - p_2)^2]$$

$$= 4\pi e^2\, (p_1 + p_2)(p_3 + p_4)[1/(p_1 - p_2)^2]$$

Later on we will discuss why we must consider only two polarizations for real photons.

Problems:

19-1. Obtain the $\pi^- - \pi^-$ scattering matrix in the c.m. system.

19-2. Do the same for the $\pi^- - \pi^+$.

19-3. Determine the Compton effect for π^+ in the frame of reference where the initial π^+ is at rest.

19-4. Calculate the $\pi^+ - \pi^-$ pair annihilation from flight with π^- at rest.

20 Virtual and Real Photons

Let us discuss the relation between virtual and real emission of photons. Why, for instance, for a real photon do we need to consider only two transverse states of polarization, while for a virtual process we have summed over all four possible states?

Suppose we send a photon to the moon. After the process is over we could describe it by a diagram, Fig. 20-1.

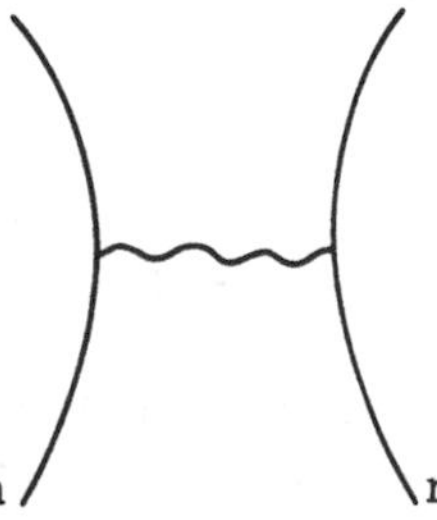

FIG. 20-1

In a sense every real photon is actually virtual if one looks over sufficiently long time scales. It is always absorbed somewhere in the universe. What characterizes a real photon is that $k^2 \to 0$ (since it is not real at all times, by the uncertainty principle, k^2 is not identically $= 0$) and therefore the propagator $1/k^2 \to \infty$. Before we proceed further with this discussion we must consider the law of conservation of charge.

Conservation of Charge. The action S for the particle plus the photon field is given by the hypothesis of minimal electromagnetic interaction:

$$S = \int d^4x\,[-\varphi^*(i\nabla_\mu - eA_\mu)^2\varphi + M^2\varphi^*\varphi + 1/4(\nabla_\nu A_\mu - \nabla_\mu A_\nu)^2]$$

$$= \int d^4x\,[-\varphi^*(i\nabla_\mu)^2\varphi + M^2\varphi^*\varphi + 1/4\,(\nabla_\nu A_\mu - \nabla_\mu A_\nu)^2]$$

$$+ eA_\mu(\varphi^*[i\nabla_\mu - (e/2)A_\mu]\varphi + \{[i\nabla_\mu - (e/2)A_\mu]\varphi\}^*\varphi)$$

Requiring that the change in the action vanish for first-order variation in the particle and photon fields we obtain the equation of motion for particle

$$(i\nabla_\mu)(i\nabla_\mu)\varphi - M^2\varphi = e[i\nabla_\mu(A_\mu\varphi) + A_\mu(i\nabla_\mu\varphi)] - e^2A_\mu A_\mu\varphi$$

and the photon

$$\nabla_\nu\nabla_\nu A_\mu = e\{\varphi^*(i\nabla_\mu - eA_\mu)\varphi + [(i\nabla_\mu - eA_\mu)\varphi]^*\varphi\}$$

(We have used the condition $\nabla_\mu A_\mu = 0$.) The charge-current vector is therefore

$$j_\mu = e\{\varphi^*(i\nabla_\mu - eA_\mu)\varphi + [(i\nabla_\mu - eA_\mu)\varphi]^*\varphi\}$$

The first-order change of action for any change in A must vanish. For the special change $\delta A_\mu = \nabla_\mu\chi$ for arbitrary χ the fields do not change, so the only change in action is that of the coupling term, or

$$\int j_\mu \nabla_\mu\chi \; d^4x = \int \chi \nabla_\mu j_\mu \; d^4x$$

Since χ is arbitrary, and this change of action vanishes, we must have

$$\nabla_\mu j_\mu = 0$$

This is the law of charge-current conservation. It is implied by the principle of gauge invariance and holds even if the hypothesis of minimal electromagnetic interactions does not.

We return now to the relation between virtual and real photons. Consider the scattering of two particles, a and b, which give rise to a current $j_\mu^a(x)$ and $j_\mu^b(x)$, respectively.

The amplitude for emitting a photon of momentum q and polarization ε is $j_\mu(q)\varepsilon_\mu$, where $j_\mu(q)$ is the Fourier transform of $j_\mu(x)$. The contribution to the scattering amplitude, due to exchange of one photon of momentum $q = (\omega, Q)$, pol ε is, according to our rules, given by

$$M = j_\mu^a(q)\,\varepsilon_\mu(1/q^2)\,j_\nu^b(q)\,\varepsilon_\nu$$

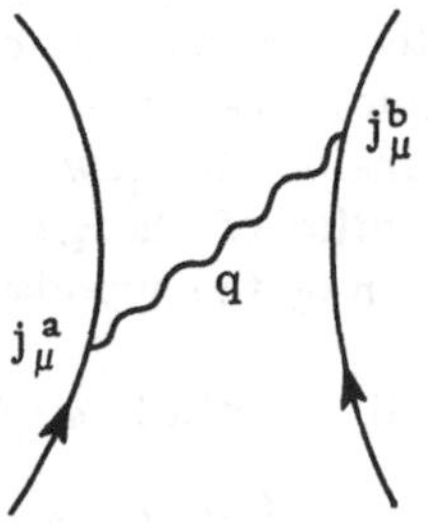

For the four possible directions of polarization of the photon we take the space-time axis, with the 3-axis oriented along the direction of propagation of the photon. On summing over polarization,

$$M = \frac{j_4^a j_4^b}{\omega^2 - Q^2} - \frac{j_3^a j_3^b}{\omega^2 - Q^2} - \frac{j_2^a j_2^b}{\omega^2 - Q^2} - \frac{j_1^a j_1^b}{\omega^2 - Q^2}$$

The last two terms are the expected contributions of the two transversely polarized photons. What is then the meaning of the first two terms? The conservation of charge current requires

$$q_\mu j_\mu (q) = 0$$

or, since the 3-axis is along Q,

$$\omega j_4 - Q j_3 = 0$$

Substituting $j_3 = (\omega/Q) j_4$ in M we find

$$M = -(j_4^a j_4^b/Q^2) - \sum_{\text{trans}} [(j^a \cdot \varepsilon)(j^b \cdot \varepsilon)/(\omega^2 - Q^2)]$$

If the photon transferred is real, $\omega \cong Q$. Then the contribution of longitudinal plus timelike photons to M (first term) vanishes, compared to that of transverse photons. However, in general, the virtual longitudinal and timelike photons cannot be neglected and, in fact, play a very important role. To see what this role is, we express the contribution of the first term in M for all momenta Q and frequency ω in coordinate space. Substituting,

$$j_4 (Q, \omega) = \int \rho(\mathbf{x}, t) \exp[-i(Q \cdot \mathbf{x} - \omega t)]\, d^3x\, dt \qquad \rho = \text{charge density}$$

we get

$$\int [j_4^a(Q,\omega)\, j_4^b(Q,\omega)/-Q^2][d^3Q\ d\omega/(2\pi)^4]$$

$$= \int \rho^a(\mathbf{x}_1 t_1)\rho^b(\mathbf{x}_2 t_2) \exp\{-i[Q(x_1 - \mathbf{x}_2) - \omega(t_1 - t_2)]\}[d^3Q\, d\omega/(2\pi)^4]$$

$$\times\ d^3\mathbf{x}_1\ dt_1\ d^3\mathbf{x}_2\ dt_2$$

The integral over ω gives $2\pi\delta(t_1 - t_2)$, and that on Q gives $4\pi/|\mathbf{x}_1 - \mathbf{x}_2|$ (since $\int \exp[-i\mathbf{Q} \cdot \mathbf{R}(d^3\mathbf{Q}/Q^2) = 4\pi/R]$ so we get

$$\int \rho^a(\mathbf{x}, t)\, \rho^b(\mathbf{x}_2\, t)/|\mathbf{x}_1 - \mathbf{x}_2|\ dt\ d^3\mathbf{x}_1\ d^3\mathbf{x}_2$$

This is the instantaneous Coulomb interaction between two charged particles.

The total interaction, which includes the interchange of transverse photons, then gives rise to the retarded interaction.

Bremsstrahlung. Suppose that a π meson scatters from a heavy particle of spin 0, for instance a K meson. Then it is possible for it to emit light. (Later we shall work out the more practical case of spin 1/2.) There are several diagrams in lowest order (Fig. 20-2) and similar diagrams, where

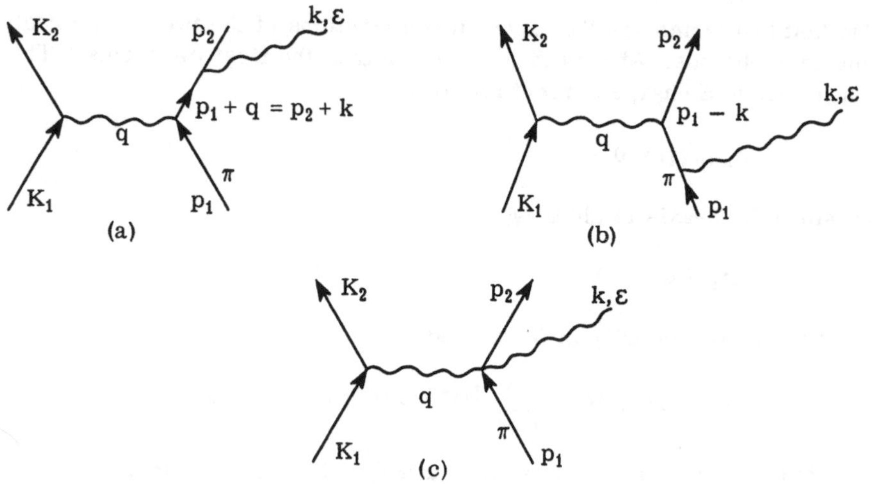

FIG. 20-2

the photon is emitted by the K. However, we are interested in a very heavy K, in which case one can show that these other diagrams can be neglected.

The amplitudes for processes (a), (b), and (c) are

$$a = [(4\pi)^{1/2}e]^3 (K_1+K_2)(2p_1+q)(2p_2+k)\cdot\varepsilon\{1/[(p_2+k)^2-M^2]\}(-1/q^2)$$

$$b = [(4\pi)^{1/2}e]^3 (K_1+K_2)(2p_2-q)(2p_1-k)\cdot\varepsilon\{1/[(p_1-k)-M^2]\}(-1/q^2)$$

$$c = (-4\pi e^2)[(4\pi)^{1/2}e](K_1+K_2)\cdot 2\varepsilon(-1/q^2)$$

Some simplifications are immediately apparent:

$$k\cdot\varepsilon = 0 \qquad (p+k)^2 - M^2 = 2p\cdot k$$

We shall consider the case in which the K meson is initially at rest and its mass $M_K \to \infty$. The conservation of energy at the photon –K meson vertex then requires that the virtual photon energy

$$\omega_q \cong Q^2/2M_K \approx 0$$

Further K_1 and K_2 have practically only time components M.

We obtain

$$a = [(4\pi)^{1/2} e]^3\, 4ME_1(p_2 \cdot \varepsilon/p_2 \cdot k)(-1/Q^2)$$

$$b = [(4\pi)^{1/2} e]^3\, 4ME_2(p_1\, \varepsilon/-p_1\, k)(-1/Q^2)$$

$$c = [(4\pi)^{1/2} e]^3\, 4M\varepsilon_4\, (-1/Q^2)$$

The heavy K meson and the π meson only exchange timelike virtual photons of zero energy. The photon propagator $1/q^2$ is then equal to $1/Q^2$, corresponding to a static Coulomb interaction. Show that the sum $a + b + c$ is gauge-invariant by showing that it vanishes if ε is in the direction k, $\varepsilon = \alpha k$. If we choose ε spacelike, diagram c vanishes.

The differential cross section for the scattering of the π meson into a solid angle $d\Omega_2$ with emission of a photon with energy ω into $d\Omega_\omega$ is (choosing ε spacelike)

$$d\sigma\, v_1 = (2\pi/2E_1\, 2M\, 2E_2\, 2M\, 2\omega)\, |a + b|^2\, D$$

where D is the density of final states (see Lecture 16):

$$D = [1/(2\pi)^6]\, E_2\, P_2\, \omega^2\, d\omega\, d\Omega\, d\Omega_\omega$$

Substituting our expression for a, b, and D in $d\sigma$ we get

$$d\sigma = \frac{4e^6}{(2\pi)^2} \frac{P_2}{P_1} \frac{\omega}{Q^4} \left| E_1 \frac{(p_2 \cdot \varepsilon)}{(p_2 \cdot k)} - E_2 \frac{(p_1 \cdot \varepsilon)}{(p_1 \cdot k)} \right|^2 d\omega\, d\Omega_2\, d\Omega_\omega$$

The conservation of total energy and momentum requires that

$$E_1 = E_2 + \omega$$

$$\mathbf{p}_1 = \mathbf{p}_2 + \mathbf{K} - \mathbf{Q}$$

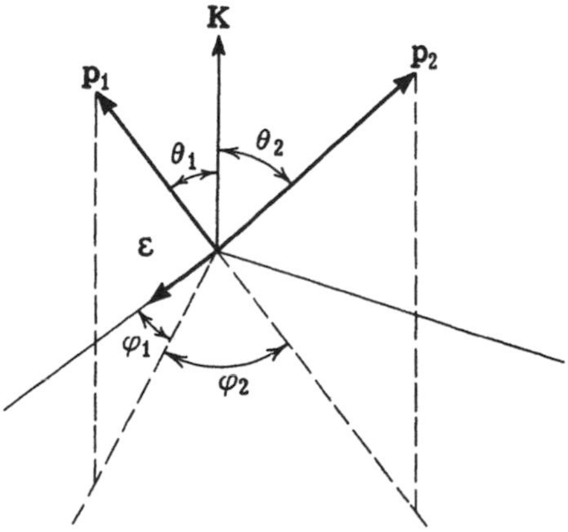

FIG. 20-3

(See Fig. 20-3.) Summing over the polarization of the emitted photons we obtain

$$d\sigma = [4e^6/(2\pi)^2](P_2/P_1)(1/Q^4)(d\omega/\omega)\, d\Omega_1\, d\Omega_2$$

$$\times\ E_1^2\left[\frac{v_2 \sin\theta_2}{1 - v_2 \cos\theta_2}\right]^2 + E_2^2\left[\frac{v_1 \sin\theta_1}{1 - v_2 \cos\theta_1}\right]^2$$

$$-\ \frac{2E_1E_2\, v_1v_2 \sin\theta_1 \sin\theta_2 \cos\varphi}{(1 - v_1 \cos\theta_1)(1 - v_2 \cos\theta_2)}$$

This is the equivalent, for particles of spin 0, of the famous Bethe-Heitler formula for particles of spin 1/2.

21 Problems

Problem 21-1: $\pi^- - \pi^-$ scattering the c.m. system. There are two diagrams:

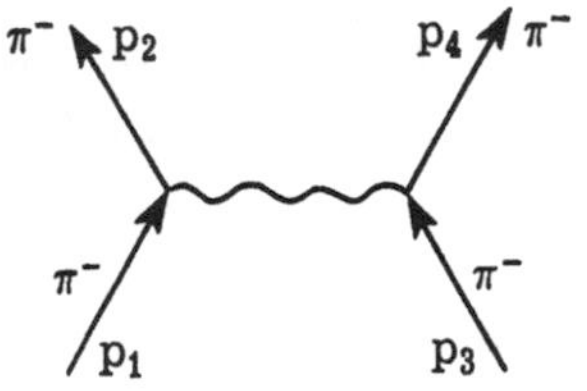

$$\text{Amp.} = [(4\pi)^{1/2}e]^2(p_1+p_2)\cdot(p_3+p_4)(1/q^2)$$

$$p_1+p_3 = p_2+p_4 \qquad q = p_1 - p_2$$

and the "exchange" diagram $p_2 \longleftrightarrow p_4$

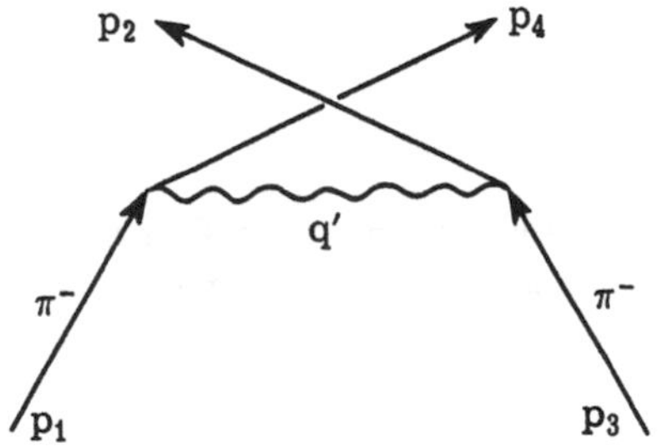

$$\text{Amp.} = [(4\pi)^{1/2}e]^2(p_1+p_4) \times (p_2+p_3)\cdot(1/q'^2)$$

$$q' = p_1 - p_4$$

In the c.m. system, $\mathbf{P}_1 = -\mathbf{P}_3 = \mathbf{P}$; $\mathbf{P}_2 = -\mathbf{P}_4 = \mathbf{Q}$; $\mathbf{P}^2 = \mathbf{Q}^2$

$$E_i = E = (\mathbf{P}^2 + M^2)^{1/2}$$

Then

$$\frac{(p_1+p_2)\cdot(p_3+p_4)}{(p_1-p_2)^2} = \frac{4E^2 + (\mathbf{P}+\mathbf{Q})^2}{(\mathbf{P}-\mathbf{Q})^2} = \frac{E^2}{P^2}\,\frac{1+v^2\cos^2\theta/2}{\sin^2\theta/2}$$

where θ is the angle between $\mathbf{P}$ and $\mathbf{Q}$, as in Fig. 21-1, and $v = P/E$.

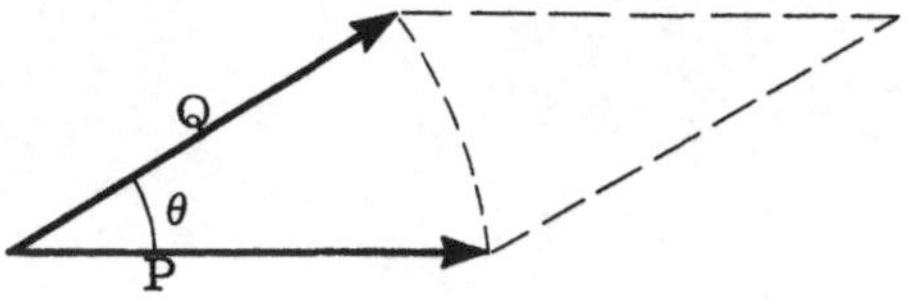

FIG. 21-1

Similarly,

$$\frac{(p_1+p_2)\cdot(P_3+P_4)}{(p_1-p_4)^2} = \frac{4E^2+(\mathbf{P}-\mathbf{Q})^2}{(\mathbf{P}+\mathbf{Q})^2} = \frac{E^2}{P^2}\,\frac{1+v^2\sin^2\theta/2}{\cos^2\theta/2}$$

Adding, we get

$$M = 4\pi e^2\,\frac{E^2}{P^2}\left[\frac{1+v^2\cos^2\theta/2}{\sin^2\theta/2} + \frac{1+v^2\sin^2\theta/2}{\cos^2\theta/2}\right]$$

Problem 21-2: $\pi^+ - \pi^-$ scattering (a very interesting case). We have a diagram (Fig. 21-2). As we discussed in Lecture 5, a π^+ (antiparticle of π^-)

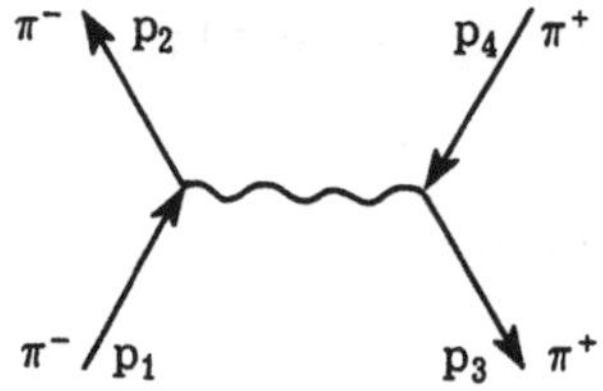

FIG. 21-2

of energy momentum P is represented by a π^- of 4-momentum $p = -P$ moving backward in time. The amplitude for the process is

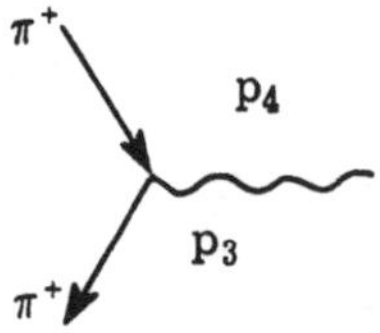

$$(4\pi)^{1/2}e\,(p_3+p_4)\cdot\mathcal{E} = -(4\pi)^{1/2}e\,(P_3+P_4)\cdot\mathcal{E}$$

which shows that the π^+ has opposite electric charge of the π^-. This is always true of charged particles and their antiparticles. The amplitude for this process is therefore

$$\text{Amp.} = -[(4\pi)^{1/2}e]^2 (p_1 + p_2) \cdot (P_3 + P_4) [1/(p_1 - p_2)^2]$$

Since the π^+ and the π^- are distinct there is, of course, no exchange diagram. However, there is an analogue to this diagram. Look at the diagram we get by changing connections so that p_1 goes to p_3 instead of p_2, and p_4 to p_2 instead of p_3.

$$\text{Amp.} = [(4\pi)^{1/2}e]^2 (p_1 - P_3) \cdot (p_2 - P_4) \times [1/(p_1 + P_3)^2]$$

in which the π^- and the π^+ annihilate, and the virtual photon recreates the pair in the final state.

We find for the scattering matrix in the c.m. system,

$$M = 4\pi e^2 \left| -\frac{E^2}{p^2} \frac{1 + v^2 \cos^2 \theta/2}{\sin^2 \theta/2} + \frac{p^2}{E^2} \cos^2 \theta \right|$$

Problem 21-3: Compton effect for π^-. We consider the process $\gamma + \pi^- \rightarrow \pi^- + \gamma$. There are three ways this can happen:

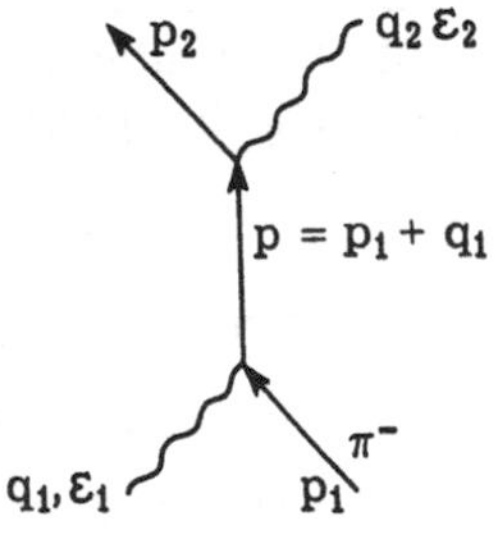

$$a = [(4\pi)^{1/2}e]^2 (2p_2 + q_2) \cdot \varepsilon_2 \times \{1/[(p_1 + q_1)^2 - m^2]\}(2p_1 + q_1) \cdot \varepsilon_1$$
$$= 4\pi e^2 [(2p_2 \cdot \varepsilon_2)(2p_1 \cdot \varepsilon_1)/2p_1 \cdot q_1]$$

since $q \cdot \varepsilon = 0$ and $(p + q)^2 - m^2 = 2p \cdot q$.

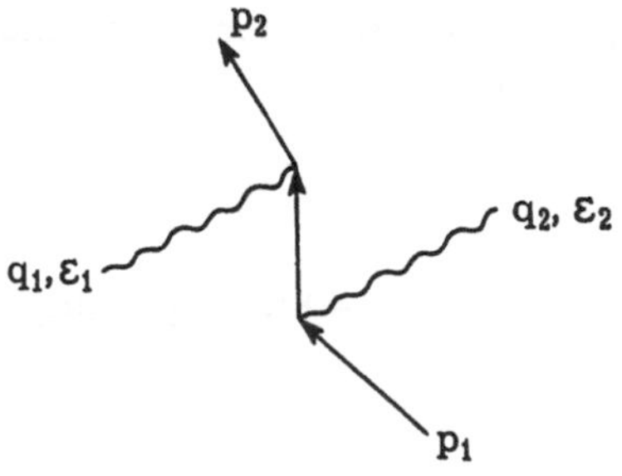

$$b = (4\pi)^{1/2}e\, (2p_2 - q_1) \cdot \varepsilon_1 \times \{1/[(p_1 - q_2)^2 - m^2]\} \times (4\pi)^{1/2}e\, (2p_1 - q_1) \cdot \varepsilon_2$$
$$= 4\pi e^2 [(2p_2 \cdot \varepsilon_1)(2p_1 \cdot \varepsilon_2)/2p_1 \cdot q$$

$$c = -4\pi e^2\, 2(\mathcal{E}_1 \cdot \mathcal{E}_2)$$

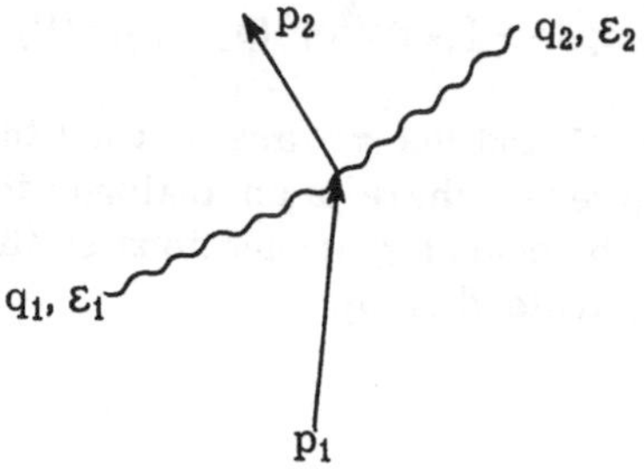

Consider the frame of reference with the initial pion at rest, where $\mathbf{P}_1 = 0$. Take $\mathcal{E}_4 = 0$. Then

$$p_1 \cdot \mathcal{E}_1 = m\, \mathcal{E}_{1_4} = 0 \qquad p_1 \cdot \mathcal{E}_2 = 0$$

and the only contribution comes from diagram c above. This result comes from our particular choice of gauge $\mathcal{E}_4 = 0$. Note that the amplitude for each diagram is not gauge-invariant. The literature is full of false remarks as to the relative magnitude of various diagrams. Only the sum is gauge-invariant. Show that the result is gauge-invariant, by showing that substituting $\mathcal{E}'_\mu = \mathcal{E}_\mu + \alpha q_\mu$ produces no change in cross section; that is, substituting $\mathcal{E}_1 = \alpha_1 q_1$ or $\mathcal{E}_2 = \alpha_2 q_2$ gives zero.

Consider the frame with the initial pion at rest, $\mathbf{P}_1 = 0$:

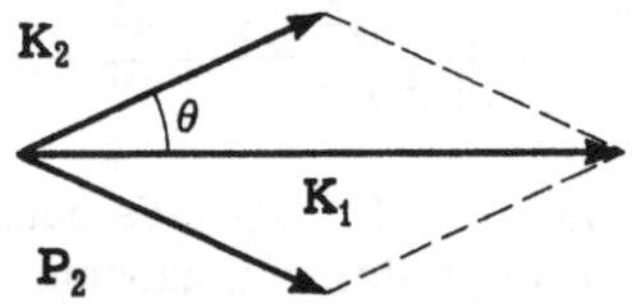

If the photon is observed, and not the pion p_2, we can get a convenient formula by eliminating p_2 from the equations by substituting $p_2 = p_1 + k_1 - k_2$. Squaring this gives $m^2 = m^2 + 2p_1 \cdot k_1 - 2p_2 \cdot k_2 - 2k_1 \cdot k_2$, or in our system,

$$m(\omega_1 - \omega_2) - \omega_1\omega_2(1 - \cos\theta) = 0$$

or the famous Compton formula,

$$1/\omega_2 = (1/\omega_1) + (1/m)(1 - \cos\theta)$$

for the change of frequency of light scattering from a free particle at rest.

Problem 21-4: $\pi^+ - \pi^-$ pair annihilation in flight. This is exactly analogous to the Compton effect, except that one of the π's is now going backward in time.

$$a = 8\pi e^2[(p_2 \cdot \mathcal{E}_2)(p_1 \cdot \mathcal{E}_1)/p_1 \cdot q_1]$$

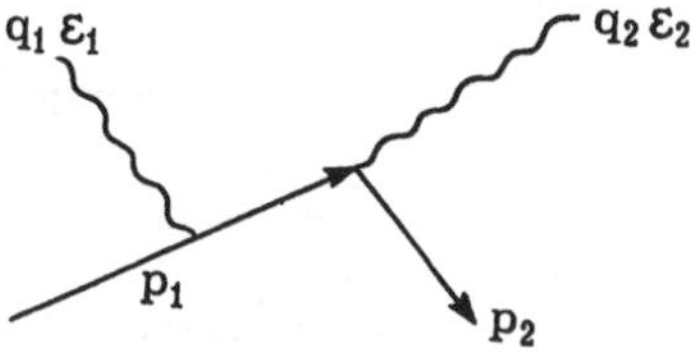

$$b = 8\pi e^2 [(p_2 \cdot \varepsilon_1)(p_1 \cdot \varepsilon_2)/p_1 \cdot q_2]$$

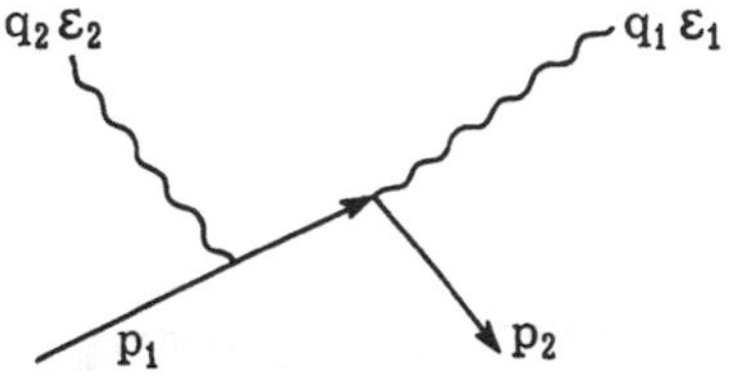

$$c = -8\pi e^2 (\varepsilon_1 \cdot \varepsilon_2)$$

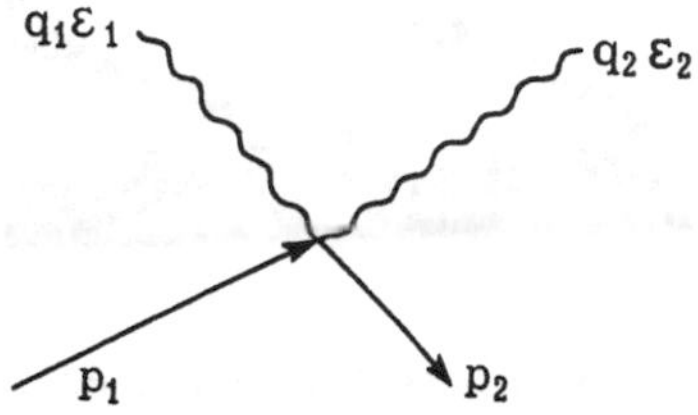

As before, we consider the frame in which $\mathbf{P}_1 = 0$. Show by squaring $\not{k}_1 = \not{p}_1 - \not{p}_2 - \not{k}_2$ that $m + E_2 = \omega_2(m + E_2 - P_2 \cos\theta)$. Then

$$M = -8\pi e^2(\varepsilon_1 \cdot \varepsilon_2)$$

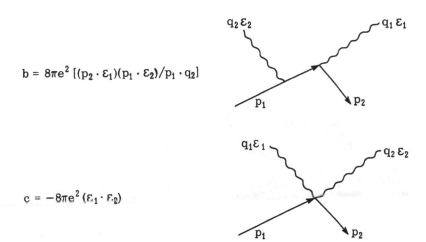

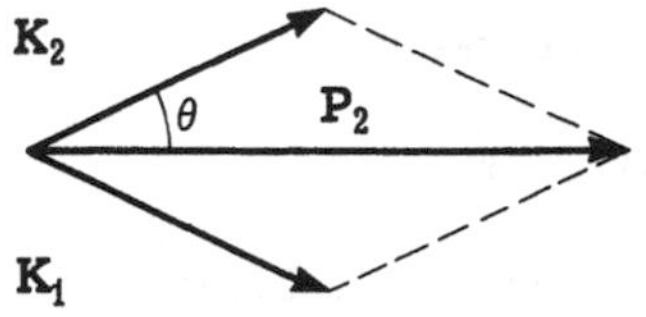

$$\mathbf{P}_2 = \mathbf{K}_1 + \mathbf{K}_2$$

$$v\, d\sigma = [e^4 \omega_2^2/E_2(m + E_2)]\, |\varepsilon_1 \cdot \varepsilon_2|^2\, d\Omega$$

We see that the total cross section is proportional to $1/v$ and becomes infinite as $v \to 0$. What does this mean? Suppose we had a gas composed of n π's per unit volume. The probability per unit time that a π^+ moving through this gas with velocity v annihilates is $1/\tau = n\sigma v$, which is a finite quantity.

For a bound $\pi^+ - \pi^-$ system (analogous to positronium) n would be equal to the square of the wave function at the origin and τ is the lifetime of the system.

22 Spin-1/2 Particles

Recall the two-component spinor, whose behavior under space rotations of angle θ about unit vector $\mathbf{n}$ was described by the operator $\exp(i\theta\mathbf{n}\cdot M)$, where $M = (1/2)\sigma$. In Problem 21-3 you were concerned with the behavior of this spinor under Lorentz transformation. As was the case in space rotations it is sufficient to consider infinitesimal transformations. We write the corresponding operator as

$$1 + i(\mathbf{v}/c)\cdot\mathbf{N}$$

where $\mathbf{v}$ is an infinitesimal velocity; $c = 1$. Proceeding as before, we have for a finite velocity v in the z direction the operator $\exp(iwN_z)$ with $\tanh w = v/c$.

Thus we need six operators to represent a general Lorentz transformation:

$$M_x \; M_y \; M_z$$
$$N_x \; N_y \; N_z$$

corresponding to the six rotations in four-dimensional space. These quantities form an antisymmetric tensor with components

$$M_{\mu\nu} = -M_{\nu\mu} = \begin{pmatrix} M_{yz} & M_{zx} & M_{zy} \\ M_{xt} & M_{yt} & M_{zt} \end{pmatrix} \qquad \text{i.e., } M_x = M_{yz}, \text{ etc.}$$

Either by algebra (studying successive Lorentz transformations) or by drawing figures we find the commutation relations

$$M_x M_z - M_y M_x = iM_z$$
$$N_x M_y - M_y N_x = iN_z$$
$$N_x N_y - N_y N_x = -M_z \qquad \text{and cyclic permutations}$$

All others commute; i.e.,

$$N_z M_z - M_z N_z = 0$$

These rules are all summarized by

$$M_{\mu\nu} M_{\sigma\tau} - M_{\sigma\tau} M_{\mu\nu} = i(\delta_{\nu\sigma} M_{\mu\tau} - \delta_{\nu\tau} M_{\mu\sigma} - \delta_{\mu\sigma} M_{\nu\tau} + \delta_{\mu\tau} M_{\nu\sigma})$$

Now we find the representation of the operator N that acts on the two-component spinor u. First of all, N must be a 2×2 matrix.

$$N_x = \begin{pmatrix} ? & ? \\ ? & ? \end{pmatrix}$$

We could put in some unknowns for the question marks and grind out the solution, using the commutation relations and $\mathbf{M} = (1/2)\boldsymbol{\sigma}$. It is easier, however, if we notice that any 2×2 matrix can be formed from a linear combination of the four matrices $1, \sigma_x, \sigma_y, \sigma_z$. So we write

$$N_x = \alpha 1 + a\sigma_x + g\sigma_y + h\sigma_z$$

We notice that N_x and σ_x commute. Therefore $g = h = 0$. We find

$$N_x = \alpha 1 + a\sigma_x$$

$$N_y = \beta 1 + b\sigma_y$$

$$N_z = \gamma 1 + c\sigma_z$$

We put these in the commutation relation

$$N_x M_y - M_y N_x = iN_z$$

$$(\alpha 1 + a\sigma_x)(1/2)\sigma_y - (1/2)\sigma_y(\alpha 1 + a\sigma_x) = i(\gamma 1 + c\sigma_z)$$

$$ia\sigma_z = i(\gamma 1 + c\sigma_z)$$

Therefore,

$$a = c$$

$$\gamma = 0$$

By cyclic permutation,

$$b = a \qquad \alpha = \beta = 0$$

Therefore,

$$\mathbf{N} = a\boldsymbol{\sigma}$$

To determine a, we substitute $\mathbf{N} = a\boldsymbol{\sigma}$ in $N_xN_y - N_yN_x = iM_z$:

$$a^2 = -1/4 \qquad a = \pm\, i/2$$

We can choose either sign for a. Suppose we choose the + sign. Then

$$\mathbf{N} = i\sigma/2 \qquad \mathbf{M} = \boldsymbol{\sigma}/2$$

However, consider the transformation properties of the mirror-image spinor. Under reflection $\mathbf{N} \to -\mathbf{N}$, since $v \to -v$ and $\mathbf{N}\cdot\mathbf{v}$ is a scalar; $\mathbf{M} \to \mathbf{M}$. Therefore a two-component spinor and its mirror image do not transform in the same way under Lorentz transformation. In order to have reflection invariance we need a four-component spinor.

By writing $\sigma_v = \boldsymbol{\sigma}\cdot\mathbf{v}/|\mathbf{v}|$ the operator transforming u under Lorentz transformation is

$$\exp(-\sigma_v w/2)$$

For instance consider the plane-wave state $u \exp(-ip\cdot x)$. For a Lorentz transformation along the z axis, $\sigma_v = \sigma_z$ and $u' = \exp(-\sigma_z w/2)u$. We can construct the general case from the transformation of $u = \begin{pmatrix}1\\0\end{pmatrix}$ and $\begin{pmatrix}0\\1\end{pmatrix}$

$$u = \begin{pmatrix}1\\0\end{pmatrix} \qquad u' = \exp(-w/2)\begin{pmatrix}1\\0\end{pmatrix}$$

$$= \begin{pmatrix}0\\1\end{pmatrix} \qquad = \exp(w/2)\begin{pmatrix}0\\1\end{pmatrix}$$

Since $\mathbf{N}$ is not Hermitian u^*u is not a scalar. Consider the transformation of u^*u:

$$u^{*\prime}u' = u^*\exp(-\sigma_z w/2)\exp(-\sigma_z w/2)u = u^*\exp(-\sigma_z w)u$$

Now,

$$\exp(-\sigma_z w) = 1 - \sigma_z w + (w^2/2!) - \sigma_z(w^3/3!) + \cdots$$

$$= [1 + (w^2/2!) + (w^4/4!) + \cdots] - \sigma_z[w + (w^3/3!) + \cdots]$$

$$\exp(-\sigma_z w) = \cosh w - \sigma_z \sinh w$$

Thus,

$$u'^* u' = \cosh w\,(u^* u) - \sinh w (u^* \sigma_z u)$$

Also,

$$u'^* \sigma_z u' = \cosh w (u^* \sigma_z u) - \sinh w (u^* u)$$

We notice immediately that $u^* u$ and $u^* \sigma_z u$ transform exactly like t, z, under Lorentz transformation:

$$\left.\begin{aligned} t' &= \gamma(t - vz) \\ z' &= \gamma(z - vt) \end{aligned}\right\} \qquad \gamma = (1 - v^2)^{-1/2}$$

Before we can conclude that $u^* u$, $u^* \sigma u$ form a 4-vector, we have to check the analogue of $x' = x$, $y' = y$:

$$\begin{aligned} u'^* \sigma_x u' &= u^* \exp(-\sigma_z w/2)\, \sigma_x \exp(-\sigma_z w/2)\, u \\ &= u^* \exp(-\sigma_z w/2) \exp(+\sigma_z w/2)\, \sigma_x u \\ &= u^* \sigma_x u \end{aligned}$$

Therefore we have discovered a new 4-vector, which we give a symbol S_μ:

$$S_\mu = u^* \sigma_\mu u$$

$$\sigma_\mu \equiv (1, \sigma)$$

Now it appears that S_μ might be satisfactory for the probability current. As before, normalize so that $u^* u = 2E$.

Then we have

$$u^* \sigma_\mu u = 2p \qquad S_\mu = 2p_\mu$$

Suppose we have a particle with spin up in the z direction:

$$u = \begin{pmatrix} 1 \\ 0 \end{pmatrix}$$

$$u^* u = 1 \qquad u^* \sigma_z u = 1 \qquad u^* \sigma_x u = u^* \sigma_y u = 0$$

There is trouble, because the probability current $u^*\sigma_z u$ is always rushing off in the z direction. This means that this probability current cannot represent a particle at rest.

Note that for this special case $(u^* u)^2 = (u^* \sigma u)^2$. This is invariant under rotations, so that, since any spinor must represent a particle spinning in some direction, calling this z, we deduce the above is true in general, That is,

$$S_\mu S_\mu = 0$$

always, or if we have $S_\mu = 2p_\mu$, we would have to have

$$p_\mu p_\mu = 0 \qquad \text{also} \qquad m = 0$$

Hence the present development is only valid for particles of 0 mass (and spin 1/2). We know of only one such particle—the neutrino. It is possible to prove in general that $S_\mu \sigma_\mu u = 0$. (Prove it by first taking the case

$$u = \begin{pmatrix} 1 \\ 0 \end{pmatrix}$$

and then arguing that it must then be true for any u.) If we take $S_\mu = 2p_\mu$ we must have

$$p_\mu \sigma_\mu u = 0$$

or

$$(E - \mathbf{p} \cdot \sigma)u = 0$$

We take this as the law describing the neutrino. It is true for each momentum plane wave, and hence for any superposition of such waves,

$$\int (E - \mathbf{p} \cdot \sigma)\, c_{\mathbf{p}} u_{\mathbf{p}} \exp(-ip \cdot x)\, [d^3\mathbf{p}/(2\pi)^3] = 0$$

where $c_{\mathbf{p}}$ is any function of momentum. We can also convert this to an equation in coordinate space.

This equation is simply

$$\int i\nabla_\mu \sigma_\mu [c_{\mathbf{p}} u_{\mathbf{p}} \exp(-ip \cdot x)]\, [d^3\mathbf{p}/(2\pi)^3] = 0$$

or

$$i\nabla_\mu \sigma_\mu \varphi(x) = 0$$

with

$$\varphi(x) = \int c_{\mathbf{p}} u_{\mathbf{p}} \exp(-ip \cdot x)\,[d^3\mathbf{p}/(2\pi)^3]$$

Written out in full, the general equation is

$$[(\partial/\partial t) + \boldsymbol{\sigma} \cdot \boldsymbol{\nabla}_-]\,\varphi(x,t) = 0$$

Define $\sigma_{\mathbf{p}} \equiv \sigma \cdot \mathbf{p}/|\mathbf{p}|$. Since $p = E$, the equation $(E - \mathbf{p} \cdot \sigma)u = 0$ is equivalent to

$$\sigma_{\mathbf{p}} u = u$$

This means that the particle always spins clockwise in the direction of motion. Actually from experiment we know that the neutrino goes counterclockwise. However, remember the other possibility for the sign of **N**.

For $\mathbf{N} = i\sigma/2$ we find that the quantity which transforms like a 4-vector is

$$S'_\mu \equiv (u^* u,\ -u^* \boldsymbol{\sigma} u)$$

In this case we get the equation

$$(E + \mathbf{p} \cdot \boldsymbol{\sigma})v = 0$$

The particle described by v spins counterclockwise:

$$\sigma_{\mathbf{p}} v = -v$$

It is essential to note that u and v transform differently:

$$u' = \exp(-\sigma_{\mathbf{v}} w/2)\,u \qquad v' = \exp(+\sigma_{\mathbf{v}} w/2)\,v$$

We say that u and v are, respectively, cospinors and contraspinors. The corresponding transformations are called covariant and contravariant.

23 Extension of Finite Mass

In Lecture 22 we saw that $S_\mu = (u^*\sigma_\mu u)$ transforms like a 4-vector. This means that, for arbitrary B_μ,

$$B_\mu (u^* \sigma_\mu u)$$

is an invariant.

From this we can see that

$$B_\mu \sigma_\mu u$$

behaves differently from u under Lorentz transformations.

Since $u' = u^* \exp(-\sigma_v w/2)$,

$$(B_\mu \sigma_\mu u)' = \exp(+\sigma_v w/2)\ (B_\mu \sigma_\mu u)$$

Thus $B_\mu \sigma_\mu u$ transforms like a contravariant spinor (u being a cospinor); if v is a contraspinor then $B_\mu \sigma_\mu v$ is a covariant spinor.

Extension to Finite Mass. We have found that for spin-1/2 particles of mass 0, the equations of motion are

$$(E - \mathbf{p} \cdot \boldsymbol{\sigma})u = 0 \qquad \text{right-handed}$$

$$(E + \mathbf{p} \cdot \boldsymbol{\sigma})v = 0 \qquad \text{left-handed}$$

We notice that such equations are not invariant under space inversion, **p** being a polar vector, $\boldsymbol{\sigma}$ an axial vector. (A few years ago this would have been sufficient reason for dropping these equations—as was done 25 years ago by Pauli in "Handbuch der Physik," p. 226—but now we know that parity is not conserved anyway, so we shall stick with our results.)

Writing the first equation in the form

$$p_\mu \sigma_\mu u = 0$$

we observe that on the left-hand side we have a contravariant quantity. Therefore if we want to add some term that describes the mass or interaction of the particle, we have to be careful that it has the same transformation property. For example, mu would be wrong, since u is a covariant spinor.

After this, the simplest eligible term for a source (linear in u) is of the form

$$A_\mu \sigma_\mu u$$

For example, the coupling for β decay recently discovered is of this form; the interaction is

$$(u_1^* \sigma_\mu u_2)(u_3^* \sigma_\mu u_4)$$

For the μ-decay, u_1 represents the neutrino, u_2 the μ meson, u_3 the electron, and u_4 the antineutrino. To find the term in the equation of motion of u_1 we vary with respect to u_1^*; noting that $u_3^* \sigma_\mu u_4$ is a vector (A_μ) we see that we get the proposed form $A_\mu \sigma_\mu u_2$.

Now consider mass again. The equation

$$(E^2 - \mathbf{p}^2)u = m^2 u$$

behaves correctly under transformation, but since u has two components, this describes two independent particles of spin 0. The real difficulty shows up when we include electromagnetic interaction. Then the above equation becomes

$$[(E - \varphi)^2 - (\mathbf{p} - \mathbf{A})^2]\, u = m^2 u$$

The term $\boldsymbol{\sigma} \cdot \mathbf{H}$ characteristic of spin-1/2 particles does not follow from this equation.

We notice further that in the absence of interaction there is no way to distinguish between

$$(E + \mathbf{p} \cdot \boldsymbol{\sigma})(E - \mathbf{p} \cdot \boldsymbol{\sigma})u = m^2 u$$

and

$$(E^2 - \mathbf{p}^2)u = m^2 u$$

However in the presence of interaction, the substitutions

$$E \rightarrow E - \varphi \qquad \mathbf{p} \rightarrow \mathbf{p} - \mathbf{A}$$

do give different results for the two equations.

Note that u is covariant; $(E - \mathbf{p}\cdot\boldsymbol{\sigma})$ makes it contravariant and $(E+\mathbf{p}\cdot\boldsymbol{\sigma})$ makes it covariant again. Thus we introduce the contravariant spinor v by

$$(E - \mathbf{p}\cdot\boldsymbol{\sigma})u = mv \tag{23-1}$$

Then we have

$$(E + \mathbf{p}\cdot\boldsymbol{\sigma})v = mu \tag{23-2}$$

These coupled equations transform correctly; when $m = 0$ they give the previous results (but are no longer coupled). Together they are equivalent to the equation

$$(E + \mathbf{p}\cdot\boldsymbol{\sigma})(E - \mathbf{p}\cdot\boldsymbol{\sigma}) = m^2u \tag{23-3}$$

We can combine (23-1) and (23-2) in a single equation by introducing the four-component spinor

$$\Psi = \begin{pmatrix} u_1 \\ u_2 \\ v_1 \\ v_2 \end{pmatrix} = \begin{pmatrix} u \\ \\ v \end{pmatrix}$$

Define the matrices

$$\gamma_t = \begin{pmatrix} 0 & 0 & 1 & 0 \\ 0 & 0 & 0 & 1 \\ 1 & 0 & 0 & 0 \\ 0 & 1 & 0 & 0 \end{pmatrix} \qquad \begin{pmatrix} 0 & 1 \\ \\ 1 & 0 \end{pmatrix}$$

$$\gamma = \begin{pmatrix} 0 & -\boldsymbol{\sigma} \\ \\ \sigma & 0 \end{pmatrix}$$

When γ_t operates on Ψ it interchanges u and v:

$$\gamma_t\Psi = \gamma_t\begin{pmatrix} u \\ v \end{pmatrix} = \begin{pmatrix} v \\ u \end{pmatrix}$$

Similarly we have

$$\gamma\begin{pmatrix} u \\ v \end{pmatrix} = \begin{pmatrix} 0 & -\sigma \\ \sigma & 0 \end{pmatrix}\begin{pmatrix} u \\ v \end{pmatrix} = \begin{pmatrix} -\sigma & v \\ \sigma & u \end{pmatrix}$$

Equations (23-1) and (23-2) are then summarized by

$$m\Psi = (E\,\gamma_t - \mathbf{p}\cdot\boldsymbol{\gamma})\Psi$$

or

$$m\Psi = p_\mu \gamma_\mu \Psi \tag{23-4}$$

Equation (23-4)[or Eqs. (23-1) and (23-2)] is known as the Dirac equation. It contains mass and has the correct transformation properties. One can think of γ_μ as behaving like a 4-vector.

[The Dirac equation is sometimes written in the form

$$(\mathbf{p}\cdot\boldsymbol{\alpha} + m\beta)\Psi = E\Psi$$

This is equivalent to (23-4) with the relations

$$\gamma_t = \beta \qquad \boldsymbol{\alpha} = \gamma_t\,\boldsymbol{\gamma}]$$

It is useful to know the properties of the γ matrices. We see easily that

$$\gamma_t^2 = 1 \qquad \gamma_x^2 = -1$$

$$\gamma_t\gamma_x + \gamma_x\gamma_t = 0$$

The complete rule is

$$\gamma_\mu\gamma_\nu + \gamma_\nu\gamma_\mu = 2\delta_{\mu\nu} \tag{23-5}$$

In most problems one need not use an explicit representation for the γ's but can derive everything from the commutation relations (23-5).

The Current. By constructing a mixture of the states u and v we can find a probability current that can also represent particles at rest. Recall that the quantities

$$(u^* u,\ u^* \boldsymbol{\sigma} u) \qquad (v^* v,\ -v^* \boldsymbol{\sigma} v)$$

are 4-vectors.

Suppose that we have a particle with spin up in the rest frame:

$$u = \begin{pmatrix}1\\0\end{pmatrix} = v, \qquad \mathbf{p} = 0, \qquad mv = Eu$$

$$(u^* u,\ u^*\boldsymbol{\sigma} u) = (1,0,0,+1)$$

$$(v^* v,\ -v^* \boldsymbol{\sigma} v) = (1,0,0,-1)$$

We note that we can cancel the space parts by defining a new 4-vector that is just the sum of the above vectors:

$$S_\mu = (u^* u + v^* v,\ u^* \sigma u - v^* \sigma v)$$

The new current has the exemplary property that its space component is 0 in the rest frame of the particle. Further simplification may be made by writing S_μ in terms of Ψ. It is easily seen that

$$S_\mu = (\Psi^* \Psi,\ \Psi^* \gamma_t \Psi)$$

where Ψ^* is the Hermitian conjugate matrix to Ψ. To put this in a more convenient form, we define

$$\bar{\Psi} \equiv \Psi^* \gamma_t$$

Then $S_\mu = (\bar{\Psi}\gamma_t \Psi, \bar{\Psi}\gamma\Psi)$ assumes the form

$$S_\mu = \bar{\Psi}\gamma_\mu \Psi \tag{23-6}$$

It is easy to see that (23-6) satisfies the continuity equation,

$$\nabla_\mu S_\mu = 0 \tag{23-7}$$

For, consider the Dirac equation and its conjugate,

$$i\nabla_\mu \gamma_\mu \Psi - m\Psi = 0$$

$$i\nabla_\mu \bar{\Psi} \gamma_\mu + m\bar{\Psi} = 0$$

Multiplying these equations by $\bar{\Psi}$ on the left and Ψ on the right, respectively, and adding, we obtain

$$i\bar{\Psi}(\nabla_\mu \gamma_\mu \Psi) + i(\nabla_\mu \bar{\Psi}\gamma_\mu)\Psi = 0$$

which is Eq. (23-7).

However u or v by themselves cannot form a conserved current. For example, $\nabla_\mu (u^* \sigma_\mu u) = 2m\ \mathrm{Im}(u^* v) \neq 0$. [This follows from Eq. (23-1) for u.]

Finally we note that Eqs. (23-1) and (23-2) are changed into each other by the transformation

$$u \to v \qquad \mathbf{p} \to -\mathbf{p}$$

S_μ is unchanged by this transformation. Thus the equations are invariant under reflection (but the β-coupling term is not).

Action Principle. The Dirac equation (23-4) [and hence (23-1) and (23-2)] may be derived from the action

$$S = \int (\bar{\Psi} p_\mu \gamma_\mu \Psi - m\bar{\Psi}\Psi)\, d^4\tau$$

Introducing the useful notation (a_μ is a 4-vector),

$$\not{a} \equiv a_\mu \gamma_\mu$$

we write the action for a particle of spin 1/2 in an electromagnetic field:

$$S' = \int [\bar{\Psi}(\not{p} - \not{A})\Psi - m\bar{\Psi}\Psi + (1/4)F_{\mu\nu}F_{\mu\nu}]\, d^4\tau$$

Varying S′ with respect to $\bar{\Psi}$ gives the equation of motion for the particle

$$(\not{p} - m)\Psi = \not{A}\Psi$$

From this equation we shall see that the propagator for a particle of spin 1/2 is $1/(\not{p} - m)$. For making calculations, since $(\not{p} - m)(\not{p} + m) = p^2 - m^2$ we shall often use the relation

$$1/(\not{p} - m) = (\not{p} + m)/(p^2 - m^2)$$

From the coupling term $e' \bar{\Psi}\not{A}\Psi$ in the Lagrangian we obtain the fundamental amplitude for the interaction of spinors and photons:

$$\text{Amp.} = (4\pi)^{1/2} e\, U_{s_2}(p_2) \not{\epsilon}\, U_{s_1}(p_1)$$

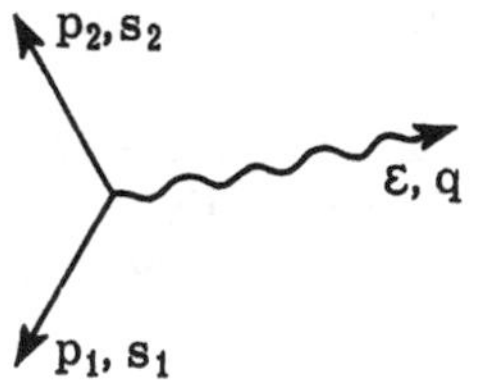

24 Properties of the Four-Component Spinor

We shall consider now the properties of the four-component spinor

$$U = \begin{pmatrix} u \\ v \end{pmatrix}$$

which satisfies the Dirac equation

$$\not{p}\, U = mU$$

or in two-component form,

$$(E - \mathbf{p} \cdot \boldsymbol{\sigma})u = mv$$

$$(E + \mathbf{p} \cdot \boldsymbol{\sigma})v = mu$$

First of all, there are only two linearly independent solutions of this equation. Consequently it represents a particle of spin 1/2. How does U transform? We have seen that under a Lorentz transformation along the z axis,

$$u' = \exp(-\sigma_z w/2)u \qquad v' = \exp(+\sigma_z w/2)v$$

Therefore,

$$U' = \begin{pmatrix} \exp(-\sigma_z w/2)u \\ \\ \exp(+\sigma_z w/2)v \end{pmatrix} = \left[\cosh(w/2) - \sinh(w/2)\begin{pmatrix} \sigma_z & 0 \\ 0 & -\sigma_z \end{pmatrix}\right] U$$

We can write this transformation in more compact form using the 4×4 matrices

$$\gamma_t = \begin{pmatrix} 0 & 1 \\ 1 & 0 \end{pmatrix} \qquad \boldsymbol{\gamma} = \begin{pmatrix} 0 & -\boldsymbol{\sigma} \\ \boldsymbol{\sigma} & 0 \end{pmatrix}$$

introduced earlier. We have

$$\gamma_t\gamma_z = \begin{pmatrix}\sigma_z & 0\\ 0 & -\sigma_z\end{pmatrix}$$

Hence

$$U' = \exp(-\gamma_t\gamma_z\ w/2)U$$

and

$$N_z = (i/2)\gamma_t\gamma_z$$

Since this transformation corresponds to a rotation in the tz plane, we expect similarly that $M_z = (i/2)\gamma_x\gamma_y$ for a rotation in the xy plane. Let us check. Substituting our representation for the γ matrices we find

$$M_z = \frac{1}{2}\begin{pmatrix}\sigma_z & 0\\ 0 & \sigma_z\end{pmatrix}$$

and

$$U' = \exp(-\gamma_x\gamma_y\ \theta_z/2)U = \begin{pmatrix}\exp(i\sigma_z\ \theta/2)u\\ \\ \exp(i\sigma_z\ \theta/2)v\end{pmatrix}$$

Let us return to the problem of describing the spin states. If the particle is at rest the Dirac equation is just

$$m\gamma_t U = mU$$

Therefore,

$$u = v$$

This shows that there are only two solutions, which we can take as spin up and down along some axis. For instance, for spin up along the z axis, we have

$$\sigma_z u = u \qquad \sigma_z v = v$$

or

$$\sigma_z U = U$$

However, if the particle is moving, $u \neq v$ (since u and v behave differently, under Lorentz transformations). We must be more careful in describing the direction of spin of a moving particle. If we take σ along the direction of motion, it is possible to describe the solutions as spin up (right helicity) or down (left helicity):

$$\not{p}U_\pm = mU_\pm \qquad \sigma_p U_\pm = \pm U_\pm$$

But σ along the direction of motion is not a Lorentz-invariant idea.

If $\boldsymbol{\sigma}$ is in an arbitrary direction, we cannot find a solution of the Dirac equation which is also an eigenfunction of $\boldsymbol{\sigma}$ ($\boldsymbol{\sigma}$ and $\not{p}$ do not commute). Let us try to find another way of describing the spin states. Returning to the rest frame we have

$$\gamma_t U = U \qquad \sigma_z U = i\gamma_x\gamma_y U = U$$

Then also

$$\sigma_z \gamma_t U = U$$

Now, we introduce the matrix

$$\gamma_5 = \gamma_t \gamma_x\gamma_y\gamma_z$$

which is Lorentz-invariant, and write

$$\sigma_z\gamma_t = i\gamma_x\gamma_y\gamma_t = i\gamma_z\gamma_5$$

Also, let W be a 4-vector satisfying $W_\mu p_\mu = 0$, $W_\mu W_\mu = -1$. In the rest frame, $W_t = 0$ and W is a unit vector in any direction. In particular if **W** is along the z axis we have $\sigma_z\gamma_t = i\not{W}\gamma_5$. Therefore U satisfies $i\not{W}\gamma_5 U = U$.

We started in the rest frame, but now the equation is Lorentz-invariant, i.e., valid in any frame. Hence for a moving particle, the 2-spin states are eigenstates of $i\not{W}\gamma_5$, where $(W \cdot p) = 0$, $W^2 = -1$. Physically they represent spin up or down along some axis in the rest frame of the particle.

When we do a problem, we shall find in general that the amplitude is of the form $m = \bar{U}_2 M U_1$, where M is a combination of γ matrices and U_1, U_2 are the initial and final spin states, respectively.

The task is to compute the probability, which is proportional to

$$\begin{aligned} m^* m &= (\bar{U}_2 M U_1)^* (\bar{U}_2 M U_1) \\ &= (\bar{U}_1 \bar{M} U_2)(\bar{U}_2 M U_1) \end{aligned}$$

where M is M with the order of all γ's reversed and each explicit $i \rightarrow -i$.

[From the definition $\bar{U} = U^* \gamma_t$ we see $\bar{M} = \gamma_t (\gamma_t M)^*$, where * means Hermitian adjoint. This rule for $\bar{M}$ does not show the invariance clearly. The rule given above is simpler. Check that they agree, for yourself. For instance,

$$\bar{\gamma}_x = \gamma_x$$

$$\overline{i\gamma_x\gamma_y} = -i\gamma_y\gamma_x$$

also

$$\bar{\gamma}_5 = \gamma_5$$

which is very useful.] There are two ways of calculating this.

The first is the obvious way. Solve the pair of equations

$$\not{p}U = mU$$

$$i\not{W}\gamma_5 U = U$$

for U_1 and U_2 and then compute

$$m = (\bar{U}_2 M U_1)$$

A much better way, which is ordinarily used in practice, is the following trick. Suppose that we are not interested in the final spin states. Then what we want is

$$\sum_{\text{2 spin states of } U_2} (\bar{U}_1 \bar{M} U_2)(\bar{U}_2 M U_1)$$

This can be written in the form

$$(\bar{U}_1 M X M U_1)$$

where $X = \Sigma_{2\ \text{spins}} U_2 \bar{U}_2$ is a 4×4 matrix (note the "wrong" order of U and $\bar{U}$). What is this matrix? Let us take a co-ordinate system in which the particle is at rest $\not{p} = m\gamma_t$. Solutions are (normalizing $\bar{U}U$ to 2m and dropping the subscript 2):

$$\text{Spin up} = (m)^{1/2}\begin{pmatrix}1\\0\\1\\0\end{pmatrix} \qquad \text{Spin down} = (m)^{1/2}\begin{pmatrix}0\\1\\0\\1\end{pmatrix}$$

Then

$$U_{up}\bar{U}_{up} = m\begin{pmatrix} 1 & 0 & 1 & 0 \\ 0 & 0 & 0 & 0 \\ 1 & 0 & 1 & 0 \\ 0 & 0 & 0 & 0 \end{pmatrix} \qquad U_{down}\bar{U}_{down} = \begin{pmatrix} 0 & 0 & 0 & 0 \\ 0 & 1 & 0 & 1 \\ 0 & 0 & 0 & 0 \\ 0 & 1 & 0 & 1 \end{pmatrix}$$

and

$$X = m\begin{pmatrix} 1 & 0 & 1 & 0 \\ 0 & 1 & 0 & 1 \\ 1 & 0 & 1 & 0 \\ 0 & 1 & 0 & 1 \end{pmatrix} = m(\gamma_t + 1)$$

or, in invariant form, $X = \not{p} + m$, which is then valid in all frames of reference.

Incidentally, another way this can be understood is to note that the law of matrix multiplication implies that

$$\sum_{\text{all 4 states of U}} (\bar{U}_1 AU)(\bar{U}BU_1) = 2m(\bar{U}_1 ABU_1)$$

The four states U are not only the two belonging to the eigenvalue $+m$ of $\not{p}$, $\not{p}U = mU$ which we want, but also two other states belonging to the other eigenvalue $-m$, $\not{p}U' = -mU'$. But if we write $A = \bar{M}(\not{p} + m)$, we shall get zero for $AU' = 0$ for the unwanted states and $AU = \bar{M}U\,2m$ for the wanted states. Therefore,

$$\sum_{\text{2 states}} (\bar{U}_1\bar{M}U_2)(\bar{U}_2MU_1) = \sum_{\text{4 states}} \{\bar{U}_1\bar{M}[(\not{p}_2 + m)/2m]U_2\}(\bar{U}_2MU_1)$$

$$= [\bar{U}_1 M(\not{p}_2 + m)MU_1]$$

If in addition the incident state is unpolarized we must average over the two spinors U_1. If we now use the fact that

$$\sum_{\text{4 states}} (\bar{U}_i AU_i) = 2m \text{ spur } A$$

we see that

$$\sum_{\text{spin 1}}\sum_{\text{spin 2}} (\bar{U}_1 MU_2)(\bar{U}_2 MU_1) = \text{spur}[\bar{M}(\not{p}_2 + m)(\not{p}_1 + m)]$$

Later on we shall discuss what to do when we are interested in the spin states.

Our whole problem has been reduced to the calculation of the spur of a combination of γ matrices. How do we calculate these spurs? We note (look at the sum of the diagonal elements of the matrices $\gamma_t, \gamma_x, \ldots$ given previously)

$$\text{sp}\,\gamma_t = 0$$

$$\text{sp}\,\gamma_x = 0$$

$$\text{sp}\,\gamma_y = 0$$

$$\text{sp}\,\gamma_z = 0$$

For any two matrices A, B,

$$\text{sp}\,AB = \text{sp}\,BA$$

$$\text{sp}(\alpha A + \beta B) = \alpha\,\text{sp}\,A + \beta\,\text{sp}\,B \qquad \alpha,\beta \text{ complex numbers}$$

Using this rule we find

$$\text{sp}\,\gamma_x\gamma_y = \text{sp}\,\gamma_y\gamma_x$$

but

$$\gamma_x\gamma_y = -\gamma_y\gamma_x$$

$$\text{sp}\,\gamma_x\gamma_y = 0$$

Only one spur is not zero—the spur of the unit matrix: spur 1 = 4.

This is a tremendous simplification; to find the spur of any complex product of γ matrices we need only find the component along the unit matrix. (There are sixteen linearly independent products of γ matrices and any 4×4 matrix can be reduced as a linear combination of these, just as any 2×2 matrix can be written as a linear combination of the three Pauli spinors σ and the identity.)

The spur of any product of an odd number of γ matrices must vanish. To reduce a product of an even number of γ matrices we proceed as follows:

$$\text{sp}\,(\not{a}\not{b}) = \text{sp}\,(\not{b}\not{a}) = (1/2)\,\text{sp}\,(\not{a}\not{b} + \not{b}\not{a})$$

$$= 4(a \cdot b)$$

$$\text{sp}\,(\not{a}\not{b}\not{c}\not{d}) = -\text{sp}\,(\not{b}\not{a}\not{c}\not{d}) + 2(a \cdot b)\,\text{sp}\,(\not{c}\not{d})$$

$$\text{sp}\,(\not{b}\not{a}\not{c}\not{d}) = -\text{sp}\,(\not{b}\not{c}\not{a}\not{d}) + 2(a \cdot c)\,\text{sp}\,(\not{b}\not{d})$$

$$\text{sp}\,(\not{b}\not{c}\not{a}\not{d}) = -\text{sp}\,(\not{b}\not{c}\not{d}\not{a}) + 2(a \cdot d)\,\text{sp}\,(\not{b}\not{c})$$

but

$$\text{sp}\,(\not{b}\not{c}\not{d}\not{a}) = \text{sp}\,(\not{a}\not{b}\not{c}\not{d})$$

$$\text{sp}\,(\not{a}\not{b}\not{c}\not{d}) = 4[(a\cdot b)(c\cdot d) - (a\cdot c)(b\cdot d) + (a\cdot d)(b\cdot c)]$$

The idea is to push the first linear combination of γ matrices to the right, at each step using the identity

$$\not{a}\not{b} = -\not{b}\not{a} + 2(a\cdot b)$$

When $\not{a}$ reaches the other side we get back the spur we started, but with opposite sign, since we have an odd number of transpositions. The remaining spurs will now consist of a product of two less γ matrices, and the whole procedure is repeated until we get to the unit matrix.

25 The Compton Effect

To get some familiarity with the spur technique we shall calculate in detail the Compton effect, the scattering of a photon from a free electron. Two diagrams contribute to this process:

$$\text{Amp.} = \bar{U}_2 (4\pi)^{1/2} e \not{\varepsilon}_2 \times [1/(\not{p}_1 + \not{q}_1 - m)] \times (4\pi)^{1/2} e \not{\varepsilon}_1 U_1$$

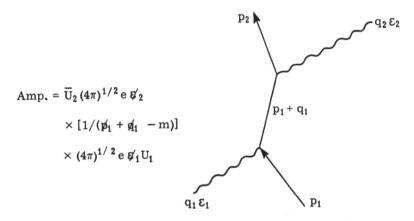

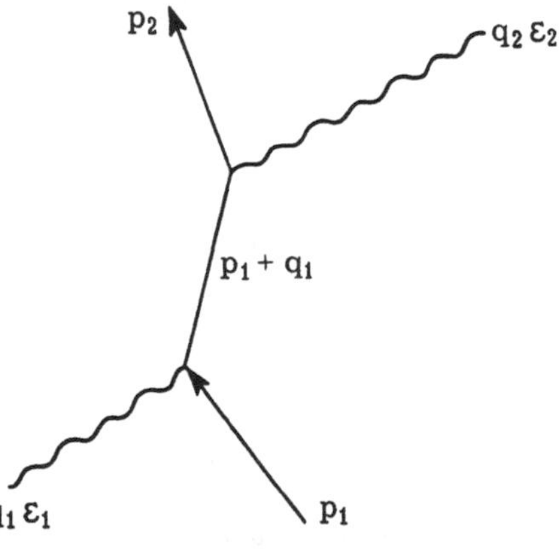

For complex polarization $\varepsilon_1, \varepsilon_2$; wave going *out* couples with ε_2^* (like an outgoing wave function). From left to right we have: $(4\pi)^{1/2} e \not{\varepsilon}_1$, amplitude to absorb the incident photon, $1/(\not{p}_1 + \not{q}_1 - m)$ = amplitude for propagation of the virtual electron, $(4\pi)^{1/2} e \not{\varepsilon}_2$ = amplitude for emission of the photon in the final state. At each vertex energy and momentum must be conserved. The total amplitude is the component between the initial and final state of the electron. Adding the diagram with the order of absorption and emission reversed we find then $m = 4\pi e^2 \bar{U}_2 M U_1$, where

$$M = \not{\varepsilon}_2 \frac{1}{\not{p}_1 + \not{q}_1 - m} \not{\varepsilon}_1 + \not{\varepsilon}_1 \frac{1}{\not{p}_1 + \not{q}_1 - m} \not{\varepsilon}_2$$

This is all the physics in the problem; the rest is pure algebra. First, we rationalize denominators,

$$\frac{1}{\not{p} - m} = \frac{1}{\not{p} - m}\,\frac{\not{p} + m}{\not{p} + m} = \frac{\not{p} + m}{p^2 - m^2}$$

Also,

$$(p_1 + q_1)^2 - m^2 = p_1^2 + 2p_1 \cdot q_1 + q_1^2 - m^2 = 2p_1 \cdot q_1$$

$$(p_1 - q_1)^2 - m^2 = -2p_1 \cdot q_2$$

and, therefore,

$$M = \not{\varepsilon}_2 \,\frac{(\not{p}_1 + \not{q}_1 + m)\not{\varepsilon}_1}{2p_1 \cdot q_1} - \not{\varepsilon}_1 \,\frac{(\not{p}_1 - \not{q}_2 + m)\not{\varepsilon}_2}{2p_1 \cdot q_2}$$

Note that M is taken between $\bar{U}_2$ and U_1, since

$$\not{p}_1 U_1 = m U_1$$

We obtain further simplification by moving $\not{p}_1$ to the right. Note that

$$\not{p}_1 \not{\varepsilon}_1 = -\not{\varepsilon}_1 \not{p}_1 + 2p_1 \cdot \varepsilon_1$$

$$\not{p}_1 \not{\varepsilon}_2 = -\not{\varepsilon}_2 \not{p}_1 + 2p_1 \cdot \varepsilon_2$$

$$M = [\not{\varepsilon}_2 q_1 \not{\varepsilon}_1 + 2\not{\varepsilon}_2 (p_1 \cdot \varepsilon_1)]/(2p_1 \cdot q_1) + [\not{\varepsilon}_1 \not{q}_2 \not{\varepsilon}_2 - 2\not{\varepsilon}_1 (p_1 \cdot \varepsilon_2)]/(2p_1 \cdot q_2)$$

Finally if we choose the frame of reference in which the electron is initially at rest, we have

$$p_1 \cdot q_1 = m\omega_1$$

$$p_1 \cdot q_2 = m\omega_2$$

$$p_1 \cdot \varepsilon_1 = p_1 \cdot \varepsilon_2 = 0$$

since ε_1 and ε_2 are spacelike and

$$M = (1/2m)[(\not{\varepsilon}_2 \not{q}_1 \not{\varepsilon}_1/\omega_1) + (\not{\varepsilon}_1 \not{q}_2 \not{\varepsilon}_2/\omega_2)]$$

To calculate the scattering cross section we need

$$1/2 \sum_{\text{spin 1}} \sum_{\text{spin 2}} (\bar{U}_1 \bar{M} U_2)(\bar{U}_2 M U_1)$$
$$= 1/2 \text{ spur } [\bar{M}(\not{p}_2 + m)\, M(\not{p}_1 + m)]$$

(see Lecture 24).

Let

$$A = \mathrm{sp}[\not\varepsilon_1^* \not q_1 \not\varepsilon_2 (\not p_2 + m) \not\varepsilon_2^* \not q_1 \not\varepsilon_1 (\not p_1 + m)]$$

$$B = \mathrm{sp}[\not\varepsilon_1^* \not q_1 \not\varepsilon_2 (\not p_2 + m) \not\varepsilon_1 \not q_2 \not\varepsilon_2^* (\not p_1 + m)]$$

$$C = \mathrm{sp}[\not\varepsilon_2 \not q_2 \not\varepsilon_1^* (\not p_2 + m) \not\varepsilon_2^* \not q_1 \not\varepsilon_1 (\not p_1 + m)]$$

$$D = \mathrm{sp}[\not\varepsilon_2 \not q_2 \not\varepsilon_1^* (\not p_2 + m) \not\varepsilon_1 \not q_2 \not\varepsilon_2^* (\not p_1 + m)]$$

Then

$$1/2\ \mathrm{sp}[\overline{M}(\not p_2 + m) M (\not p_1 + m)] = (1/8m^2)[(1/\omega_1^2)A + (1/\omega_1\omega_2)(B + C) + (1/\omega_2^2)D]$$

Consider A. Since $\not\varepsilon_1 \not\varepsilon_1^* = -1$, we shall try to get the two $\not\varepsilon_1$'s together. Note that $\not p_1 \not\varepsilon_1 = -\not\varepsilon_1 \not p_1$.

$$A = \mathrm{sp}[\not\varepsilon_1^* \not q_1 \not\varepsilon_2 (\not p_2 + m) \not\varepsilon_2^* \not q_1 \not\varepsilon_2^* \not q_1 \not\varepsilon_1 (\not p_1 + m)]$$

$$= \mathrm{sp}[\not q_1 \not\varepsilon_2 (\not p_2 + m) \not\varepsilon_2^* \not q_1 (\not p_1 - m)]$$

$$= 2(p_2 \cdot \varepsilon_2^*)\ \mathrm{sp}[\not q_1 \not\varepsilon_2 \not q_1 (\not p_1 - m)] + \mathrm{sp}[\not q_1 (\not p_2 - m) \not q_1 (\not p_1 - m)]$$

Now we use

$$\mathrm{sp}\ \not a \not b = 4 a \cdot b \qquad \mathrm{sp}\ \not a \not b \not c \not d = 4[(a \cdot b)(c \cdot d) - (a \cdot c)(b \cdot d) + (a \cdot d)(b \cdot c)]$$

to get

$$A = 8[2(p_2 \cdot \varepsilon_2^*)(q_1 \cdot \varepsilon_2) + (q_1 \cdot p_2)](q_1 \cdot p_1)$$

If we interchange $\varepsilon_1 \leftrightarrow \varepsilon_2$, $q_1 \leftrightarrow q_2$ we get

$$D = 8[2(p_2 \cdot \varepsilon_1)(q_2 \cdot \varepsilon_1) + (q_2 \cdot p_2)](q_2 \cdot p_1)$$

For $p_1 = (m, 0)$

$$p_2 \cdot \varepsilon_2 = q_1 \cdot \varepsilon_2 \qquad p_2 \cdot \varepsilon_1 = -q_2 \cdot \varepsilon_1$$

$$q_1 \cdot p_2 = m\omega_2 \qquad q_2 \cdot p_2 = m\omega_1$$

and

$$A = 8m\,\omega_1[2|(q_1\cdot\varepsilon_2)|^2 + m\omega_2]$$

$$D = 8m\omega_2[-2|(q_2\cdot\varepsilon_1)|^2 + m\omega_1]$$

Next consider B. We begin by moving $\not{\varepsilon}_1$ to the right:

$$B = \mathrm{sp}[\not{\varepsilon}_1^*\not{q}_1\not{\varepsilon}_2(\not{p}_2 + m)\not{\varepsilon}_1\not{q}_2\not{\varepsilon}_2^*(\not{p}_1 + m)]$$

$$= 2(\varepsilon_1\cdot q_2)\{\mathrm{sp}[\not{\varepsilon}_1^*\not{q}_1\not{\varepsilon}_2(\not{p}_2 + m)\not{\varepsilon}_2^*(\not{p}_1 + m)] = \alpha\}$$

$$- 2(\varepsilon_1\cdot\varepsilon_2^*)\{\mathrm{sp}[\not{\varepsilon}_1^*\not{q}_1\not{\varepsilon}_2(\not{p}_2 + m)\not{q}_2(\not{p}_1 + m)] = \beta\}$$

$$+\{\mathrm{sp}[\not{q}_1\not{\varepsilon}_2(\not{p}_2 + m)\not{q}_2\not{\varepsilon}_2^*(\not{p}_1 - m)] = \gamma\}$$

Pushing $\not{\varepsilon}_2$ to the right in α and γ and substituting $\not{p}_2 = \not{p}_1 + \not{q}_1 - \not{q}_2$ in β we get

$$\alpha = 2(p_2\cdot\varepsilon_2^*)\,\mathrm{sp}[\not{\varepsilon}_1^*\not{q}_1\not{\varepsilon}_2(\not{p}_1 + m)] - \mathrm{sp}[\not{\varepsilon}_1^*\not{q}_1(\not{p}_2 - m)(\not{p}_1 + m)]$$

$$\beta = \mathrm{sp}[\not{\varepsilon}_1^*\not{q}_1\not{\varepsilon}_2(\not{p}_1 + m)\not{q}_2(\not{p}_1 + m)] + \mathrm{sp}[\not{\varepsilon}_1^*\not{q}_1\not{\varepsilon}_2\not{q}_1\not{q}_2(\not{p}_1 + m)]$$

$$= 2(p_1\cdot q_2)\,\mathrm{sp}[\not{\varepsilon}_1^*\not{q}_1\not{\varepsilon}_2(\not{p}_1 + m)] + 2(q_1\cdot\varepsilon_2)\,\mathrm{sp}[\not{\varepsilon}_1^*\not{q}_1\not{q}_2(\not{p}_1 + m)]$$

$$\gamma = 2(p_2\cdot\varepsilon_2)\,\mathrm{sp}[\not{q}_1\not{q}_2\not{\varepsilon}_2^*(\not{p}_1 - m)] - \mathrm{sp}[\not{q}_1(\not{p}_2 - m)\not{q}_2(\not{p}_1 - m)]$$

Evaluating the spurs we get

$$\alpha = 4\{2(\varepsilon_2^*\cdot p_2)[-(\varepsilon_1^*\cdot\varepsilon_2)(q_1\cdot p_1) + (\varepsilon_1^*\cdot p_1)(q_1\cdot\varepsilon_2)] - (\varepsilon_1^*\cdot p_2)$$

$$\times (q_1\cdot p_1) + (\varepsilon_1^*\cdot p_1)(q_1\cdot p_2)\}$$

$$= 4\{2(\varepsilon_2^*\cdot q_1)[-(\varepsilon_1^*\cdot\varepsilon_2)] + (\varepsilon_1^*\cdot q_2)\}\ m\omega_1$$

$$\beta = 4\{2(p_1\cdot q_2)[-(\varepsilon_1^*\cdot\varepsilon_2)(q_1\cdot p_1) + (\varepsilon_1^*\cdot p_1)(q_1\cdot\varepsilon_2)]$$

$$+ 2(q_1\cdot\varepsilon_2)[-(\varepsilon_1^*\cdot q_2)(q_1\cdot p_1) + (\varepsilon_1^*\cdot p_1)(q_1\cdot q_2)]\}$$

$$= 4\{2m\omega_2(-\varepsilon_1\cdot\varepsilon_2^*) + 2(q_1\cdot\varepsilon_2)[-(\varepsilon_1^*\cdot q_2)]\}\ m\omega_1$$

$$\gamma = 4\{2(p_2\cdot\varepsilon_2)[(q_1\cdot q_2)(\varepsilon_2^*\cdot p_1) - (q_1\cdot\varepsilon_2^*)(q_2\cdot p_1)] - (q_1\cdot p_2)(q_2\cdot p_1)$$

$$+ (q_1\cdot q_2)(p_2\cdot p_1) - (q_1\cdot p_1)(p_2\cdot q_2) - m^2(q_1\cdot q_2)\}$$

$$= 4\{2(q_1 \cdot \varepsilon_2^*)(-q_1 \cdot \varepsilon_2)\,\omega_2 m - (q_1 \cdot p_2)\,\omega_2 m + (q_1 \cdot q_2)(p_1 \cdot p_2)$$

$$- \omega_1 m(p_2 \cdot q_2) - m^2 q_1 \cdot q_2\}$$

$$= 4\{2(q_1 \cdot \varepsilon_2^*)(-q_1 \cdot \varepsilon_2)\,\omega_2 m - (\omega_1^2 + \omega_2^2)m + (q_1 \cdot q_2)(p_1 \cdot p_2 - m^2)\}$$

The last term can be simplified by substituting

$$(q_1 \cdot q_2) = (1/2)(q_1 - q_2)^2 = -(1/2)(p_1 - p_2)^2 + p_1 \cdot p_2 - m^2$$

$$= m(\omega_1 - \omega_2)$$

$$\gamma = 4\{2(q_1 \cdot \varepsilon_2^*)(-q_1 \cdot \varepsilon_2)\,\omega_2 - 2\omega_1\omega_2\}\ m$$

Finally,

$$B = 8\{|\varepsilon_1 \cdot q_2|^2\, m\omega - |q_1 \cdot \varepsilon_2|^2\, m\omega_2 + m\omega_1 m\omega_2\, [2|\varepsilon_1 \cdot \varepsilon_2^*|^2 - 1]$$

$$+ [2(\varepsilon_1^* \cdot \varepsilon_2)(\varepsilon_2^* \cdot q_1)(\varepsilon_1 \cdot q_2) - 2(\varepsilon_1 \cdot \varepsilon_2^*)(\varepsilon_2 \cdot q_1)(\varepsilon_1^* \cdot q_2)]\ m\omega_1\}$$

A similar calculation for C gives $B^* = C$ (so that the last two terms in B cancel out in the sum B + C). Note that this result cannot be obtained by just interchanging $\varepsilon_1 \leftrightarrow \varepsilon_2$ and $q_1 \leftrightarrow q_2$ in the final expression for B, because we obtained it in a special frame of reference; namely, that one for which $p_1 = (m, 0)$. (One gets it by reversing the order of all factors in B.) Collecting our results we get

$$1/2\ \text{spur}\,[\overline{M}(\not{p}_2 + m)(M(\not{p}_1 + m)]$$

$$= (1/m^2)[(m/\omega_1)(2|q_1 \cdot \varepsilon_2|^2 + m\omega_2) + (m/\omega_2)(-2|q_2 \cdot \varepsilon_1|^2 + m\omega_1)]$$

$$+ (2/\omega_1\omega_2)[|\varepsilon_1 \cdot q_2|^2\, m\omega_1 - |q_1 \cdot \varepsilon_2|^2\, m\omega_2 + m\omega_1 m\omega_2$$

$$\times (2|\varepsilon_1 \cdot \varepsilon_2|^2 - 1)]$$

$$= [(\omega_1/\omega_2) + (\omega_2/\omega_1) - 2 + 4(\varepsilon_1 \cdot \varepsilon_2^*)(\varepsilon_1^* \cdot \varepsilon_2)|\,]$$

The scattering cross section is given by (see Lecture 16)

$$d\sigma = [(4\pi)^2 e^4/2^4 m\omega_1 E_2\,\omega_2]\,[(\omega_1/\omega_2) + (\omega_2/\omega_1) - 2 + 4|\varepsilon_1\ \ \varepsilon_2^*|^2]$$

$$\times\ 2\pi D$$

D is the density of outgoing states per unit range

$$D = E_2\omega_2^3\, d\Omega/(2\pi)^3 m\omega_1 \qquad E_2 = m + \omega_1 - \omega_2$$

$$1/\omega_2 = (1/\omega_1) + (1/m)(1 - \cos\theta)$$

θ is the angle between the incident and outgoing photon. Substituting in $d\sigma$ we obtain finally

$$d\sigma/d\Omega = (r_0^2/4)(\omega_2/\omega_1)^2[(\omega_1/\omega_2) + (\omega_2/\omega_1) - 2 + 4|\varepsilon_1 \cdot \varepsilon_2^*|^2]$$

$$r_0 = e^2/m$$

In the nonrelativistic limit ($\omega_1 \ll m$) we have $\omega_1 \approx \omega_2$ and the cross section reduces to that for a scalar particle

$$d\sigma/d\Omega = r_0^2(\omega_2/\omega_1)^2|\varepsilon_1 \cdot \varepsilon_2^*|^2$$

while in the extreme relativistic limit ($\omega_1 \gg m$) we have $\omega_2 \approx m \ll \omega_1$ (except near $\theta = 0$), and

$$d\sigma/d\Omega = (r_0^2/4)(\omega_2/\omega_1)$$

Physically this means that in the nonrelativistic limit the interaction takes place mainly through the charge, whereas in the relativistic limit it is through the magnetic moment of the electron.

26 Direct Pair Production by Muons

As another example consider direct pair production by a muon incident on a very heavy nucleus of mass M, spin zero, and charge Ze. In the laboratory the process is something like the one shown in Fig. 26-1. The im-

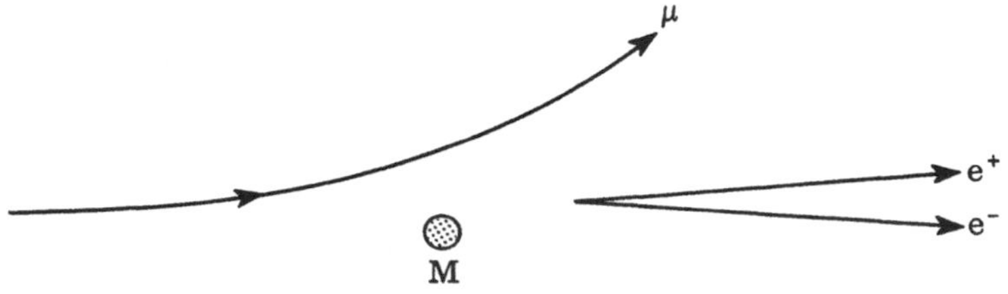

FIG. 26-1

portant processes contributing to direct pair production are shown in Fig. 26-2 (a) and (b). Note that $-q_1$ is the true momentum of the positron. From the conservation laws:

$$k_1 = P_1 - P_2$$

$$k_2 = p_1 - p_2$$

$$q_1 + p_1 + P_1 = q_2 + P_2 + p_2$$

There are also two other ways for direct pair production to occur. We just state that they are negligible for the case considered, Fig. 26-2 (c) and (d). The essential reason that diagrams (c) and (d) are negligible is that it is difficult for a heavy particle to emit a photon. For direct pair production by electrons this argument does not apply, so that diagram (c) will become important.

For diagram (a) the amplitude is

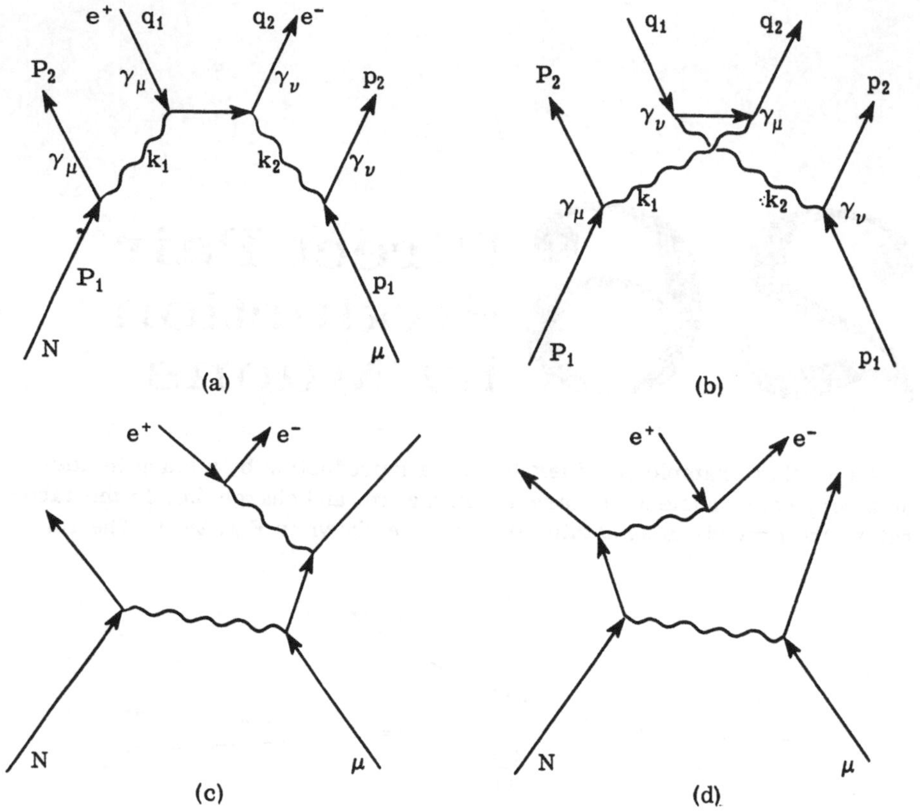

FIG. 26-2

$$4\pi e^2 Z\{\bar{U}(q_2)\gamma_\nu [1/(\not{q}_1 + \not{k}_1 - m)]\gamma_\mu U(q_1)\}(1/k_1^2)(1/k_2^2)$$

$$\times [\bar{U}(p_2)\gamma_\nu U(p_1)](P_{1_\mu} + P_{2_\mu})$$

In deriving this result we have followed each particle along its world line. $U(p_1)$, $U(p_2)$ refer to the muon states; $U(q_1)$, $U(q_2)$ refer to the electron states; m is the electron mass.

The amplitude corresponding to diagram (b) is slightly different. It is

$$4\pi e^2 Z\{\bar{U}(q_2)\gamma_\mu [1/(\not{q}_1 + \not{k}_1 - m)]\gamma_\nu U(q_1)\}(1/k_1^2)(1/k_2^2)$$

$$\times [\bar{U}(P_2)\gamma_\nu U(p_1)](P_{1_\mu} + P_{2_\mu})$$

Now if M is very large

$$P_{1_\mu} = P_{2_\mu} = M\delta_{\mu_4}$$

This approximation corresponds to the neglect of the recoil of the nucleus. We now show that neglecting recoil is equivalent to including only the Coulomb interaction with the nucleus. Suppose the nucleus is initially at rest. Since $P_2 = P_1 + k_1$, we have

$$2P_1 \cdot k_1 + k_1^2 = 0$$

or

$$2M\omega_1 + \omega_1^2 - \mathbf{K}_1^2 = 0$$

so that

$$\omega_1 \cong \mathbf{K}_1^2/2M = 0$$

Hence

$$1/k_1^2 \cong -1/\mathbf{K}_1^2$$

The latter form is just the Coulomb potential in momentum space.

One can take into account the screening of the nucleus by surrounding electrons, for example, by replacing $1/\mathbf{K}_1^2$ by the Fourier transform of an effective potential. For instance, if $V(r) = (Ze^2/r)\exp(-\alpha r)$, the appropriate form is

$$1/(\mathbf{K}_1^2 + \alpha^2)$$

We have assumed that the spin of the nucleus is 0 [amplitude for emitting a photon $(4\pi)^{1/2} Ze(P_1 + P_2) \cdot \varepsilon$]. Suppose the nucleus had spin 1/2. Then the amplitude for emitting a photon is $(4\pi)^{1/2} Ze\bar{U}(P_2)\not{\varepsilon}(P_1)$. We have

$$\bar{U}(P_2)\gamma_\mu U(P_1) = (1/2M)\bar{U}(P_2)(\not{P}_2\gamma_\mu + \gamma_\mu \not{P}_1)U(P_1)$$

Substituting $k_1 = P_2 - P_1$, we get

$$\bar{U}(P_2)\gamma_\mu U(P_1) = (1/2M)(P_{1\mu} + P_{2\mu})\,\bar{U}(P_2)U(P_1) + (1/4M)\bar{U}(P_2)$$
$$\times (\not{k}_1\gamma_\mu - \gamma_\mu \not{k}_1)\,U(P_1)$$

The first term is the contribution of the charge. In the limit $M \to \infty$, $\bar{U}(P_2)U(P_1) = 2M$ (no spin flip) and we get the same result as in the 0-spin case. The second term is the contribution of the magnetic moment. It is proportional to the recoil momentum, which is negligible in this case.

27 Higher-Order Processes

Consider the scattering of two electrons. The lowest-order term corresponded to the diagram

$$\text{Amp.} = [(4\pi)^{1/2}e]^2 (\bar{U}_4\gamma_\mu U_3)(\bar{U}_2\gamma_\mu U_1) \times (1/q^2)$$

$$q = p_2 - p_1$$

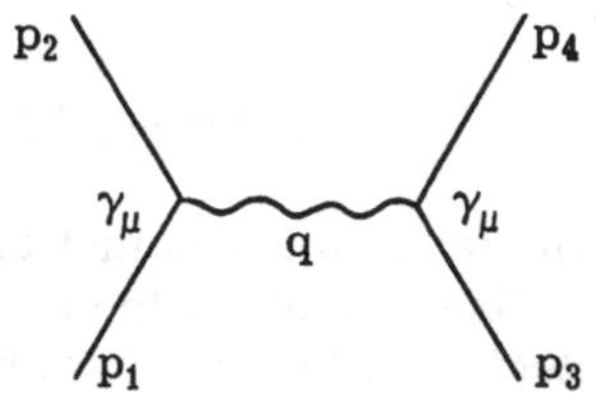

and the "exchange" diagram.

Now suppose we want to know the answer more accurately. It is then necessary to consider the two diagrams of Fig. 27-1.

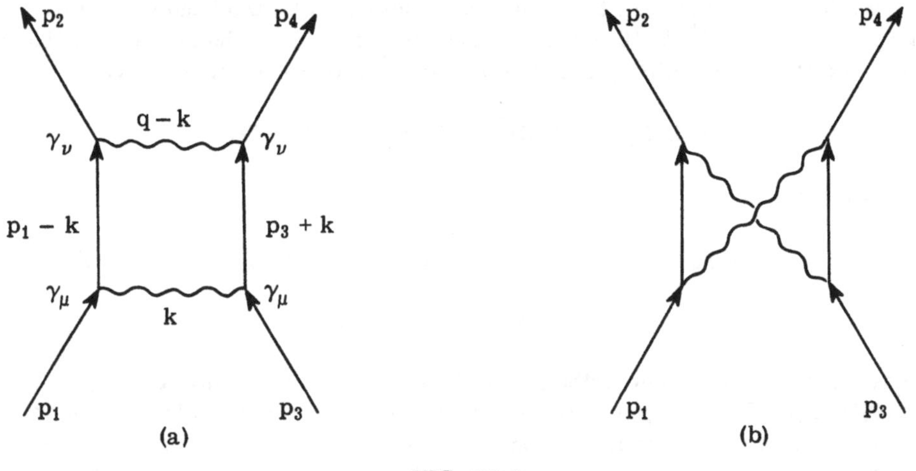

FIG. 27-1

The amplitude for diagram (a) is

$$[(4\pi)^{1/2}e]^4 \int \{\bar{U}_4 \gamma_\nu [1/(\not{p}_3 + \not{k} - m)] \gamma_\mu U_3\} \{\bar{U}_2 \gamma_\nu [1/(\not{p}_1 - \not{k} - m)] \gamma_\mu U_1\}$$

$$\times (1/k^2)[1/(q-k)^2] [d^4 k/(2\pi)^4]$$

Note that we have integrated the amplitude over all possible photon momenta in the intermediate states. The factors $1/k^2$, $1/(q-k)^2$ correspond to the propagation of the photons. The convention used in writing the factors in parentheses is that right to left in the matrix element corresponds to following the arrow attached to the line of each external particle. Each external particle line is being followed independently of the others. Thus the factor $\{\bar{U}_2 \gamma_\nu [1/(\not{p}_1 - \not{k} - m)] \gamma_\mu U_1\}$ says that a particle in state U_1 emits a photon of polarization μ (factor γ_μ), momentum k; after propagating according to $1/(\not{p}_1 - \not{k} - m)$ it is scattered to the state U_2 by interacting with another virtual photon (γ_ν).

This integral can be computed. It behaves at large k like $\int^\infty (k^3\, dk/k^6)$ so it is convergent. However, there is some trouble at small k, but this has a good physical explanation. It turns out that we are asking the wrong question when we demand the amplitude for scattering when *no* photon is emitted, as is implicitly implied by the diagrams on p. 134. In fact we cannot scatter two electrons without some low-energy photon being emitted. Instead we must ask for the probability that no radiation energy greater than $\Delta\varepsilon$ be omitted. This is equal to the probability that no photon at all be emitted plus the probability of one photon of $E_1 < \Delta\varepsilon$ plus probability two photons with total energy $E_2 < \Delta\varepsilon$, etc. The first two terms are separately divergent in order e^6, but if we add them together we get a finite answer, in order e^6.

For instance for small ω the photon-emission amplitude varies like $\int (d\omega/\omega)$. If we cut off the lower limit at x, and similarly for the amplitudes of diagrams (a) and (b), then x will cancel out of the result, so that the limit $x \to 0$ may be taken. However this direct procedure is not easily maintained relativistically invariant and can thereby cause trouble.

Instead suppose that the photon had a very small mass (mass is invariant) λ. Thus in the photon propagator we replace k^2 by $k^2 - \lambda^2$. Then the amplitude for diagrams (a) + (b) contains a term $\ln(m/\lambda)$.

Now consider the lowest-order amplitude diagrams (a) + (b). The cross term is proportional to e^6.

The probability for no photon is then proportional to $e^4 + ae^6 \ln(m/\lambda)$, where a is some number greater than 0.

Now for the diagram (e^6) of Fig. 27-2. We find that the probability of one photon with $E < \Delta\varepsilon$ is

$$-e^6 a \ln(\Delta\varepsilon/\lambda)$$

The other numerical factors are the same.

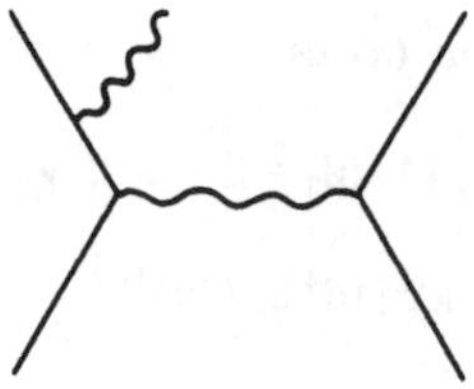

FIG. 27-2

Thus, adding, λ goes away to the order (e^6). All such divergences (called the infrared catastrophe) go away if the proper question is asked. Bloch and Nordsieck first saw the answer to this problem.

It might be objected that as $\lambda \to 0$ and $e^2 \ln(m/\lambda) \to \infty$, the perturbation theory is not valid. Yet we have a lot of room in which to work in practice. We might ask how small λ would have to be before the correction is not small. Thus we require that

$$e^2 \ln(m/\lambda) \ll 1$$

or

$$\lambda/m \ll e^{-137} \approx 10^{-60}$$

$$\lambda \ll 10^{-60}\, m$$

We have seen that the so-called infrared catastrophe was really no catastrophe at all.

Next, we turn to an entirely new type of diagram. A photon emitted by an electron may be absorbed by the *same* electron.

An example of such a process is the diagram of Fig. 27-3.

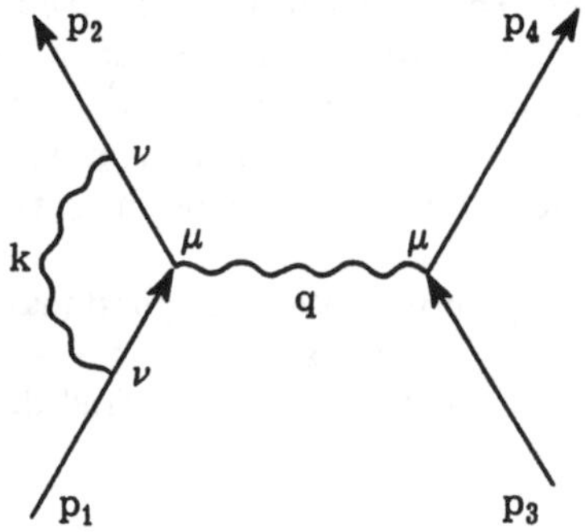

FIG. 27-3

The amplitude is

$$[(4\pi)^{1/2}e]^4 \int (\bar{U}_4\gamma_\mu U_3)\{\bar{U}_2\gamma_\nu[1/(\not{p}_2-\not{k}-m)]\gamma_\mu[1/(\not{p}_1-\not{k}-m)]\gamma_\nu U_1\}$$
$$\times (1/q^2)(1/k^2)[d^4k/(2\pi)^4]$$

Now we have a disease—trouble! For large k this integral goes like

$$\int k^3\,dk/k^4$$

since strictly we should integrate over *all* k, and the integral is logarithmically divergent for large k.

This is called the ultraviolet catastrophe, in order to distinguish it from the other. In contrast to the previous case this really is a catastrophe. It is *not* solved. But we do have a method of sweeping the dirt under the rug: First, let us write down all the fourth-order diagrams (Fig. 27-4). There is one more, corresponding to another disease known as vacuum polarization (Fig. 27-5).

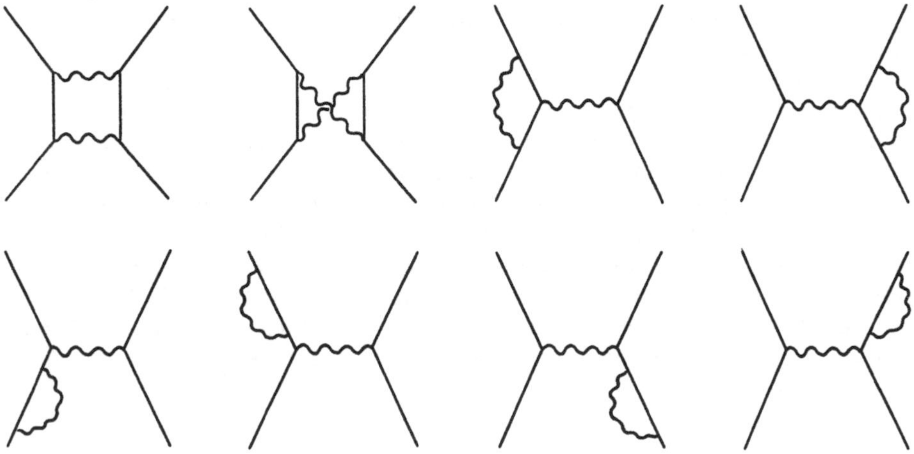

FIG. 27-4

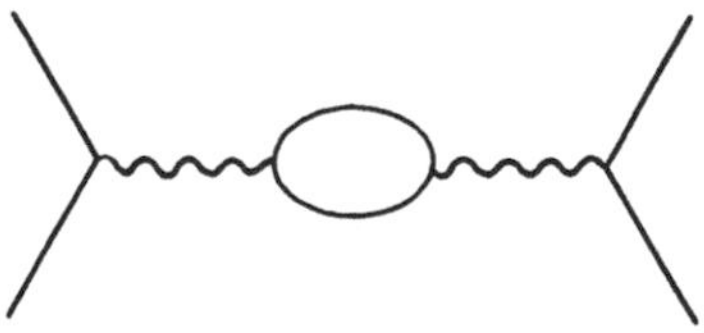

FIG. 27-5

Before attacking these problems consider a simple case. In second order we can have the diagram of Fig. 27-6, corresponding to virtual emission and absorption of the photon by the electron.

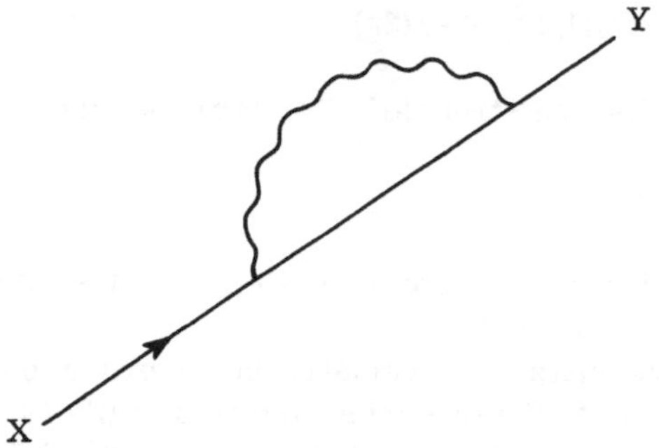

FIG. 27-6

Now nothing is really free. For an electron going from X to Y, the pole of the propagator *for a free particle* is at $p^2 = m^2$. However, making measurements at X and Y we could not tell if the electron had emitted and absorbed any number of photons. Such processes, the simplest of which is shown above, cause a shift in the position of the pole. Physically, this means that what we measure (the "experimental" mass, m_{exp}) is not the "bare" mass but something else which includes the effect of the virtual processes mentioned above. Mathematically we define the experimental mass as the location of the pole of the propagator when the above processes are taken into account. This discussion shows that the "bare" mass (which will now be called m_0) is in fact not directly observable. By using this fact we can invent a prescription that circumscribes (but does not "solve") the divergence difficulties in electrodynamics. This procedure does not work for all the arbitrary theories that people have invented; for example, the pseudovector meson theory.

28 Self-Energy of the Electron

The self-energy of the electron is an old problem; it appeared in classical physics. If you assume that an electron is a ball of radius a with all its charge on the surface, then the total electrostatic energy $E_0 = e^2/2a$. Maybe the mass m of the electron corresponds to this energy. However, if you compute the momentum P carried by the field when the electron is moving with velocity V (including the Lorentz contraction of the ball) you find $P = (4/3)E_0 V/(1 - V^2)^{1/2}$. This corresponds to a particle of mass $m = (2/3) \times (e^2/a)$. Poincaré suggested that something must hold the ball together and that the forces would contribute an additional amount of stored energy. But there is no reliable theory for these forces.

This self-energy comes from the energy needed to "assemble" the charge. From one view it is the energy of interaction of one part of the electron charge with another. One way out might at first seem to be to deny that an electron can act on itself—to suppose that electrons act only on each other. (Then the electron could be a point charge.) Yet the action of an electron on itself is required to explain a real phenomena, that of radiation resistance. An accelerating charge radiates, losing energy, so the accelerating force must do work. Against what? Against a force generated by the action of one part of the charge on another, according to classical physics.

You can calculate the force F on a moving charge ball due to the interaction of the electromagnetic field of one part of the ball on another. This force is

$$F = (2/3)(e^2/a)\ddot{\chi} + (2/3)(e^2/c^3)\dddot{\chi} + 0(a)$$

The first term agrees with the mass computed from the momentum of the field. The second term is the reaction force due to the radiation emitted by the electron and does not depend on a. However, it is not consistent to let $a \rightarrow 0$. A spread-out charge has never been thoroughly analyzed. Problems about internal motions, etc., arise. Actually, these problems were solved classically in various ways, but none of the ways have been carried over successfully in quantum mechanics. (For references see R. P. Feynman.[10])

Mass Renormalization. We discuss now the analogue of this problem in quantum mechanics, mass renormalization. Consider the amplitude for an electron to propagate between two vertices X and Y. The lowest-order diagram is

$$\text{Amp.} = Y \frac{1}{\not{p} - m} X$$

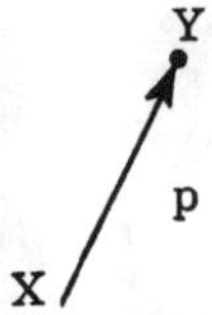

It is also possible that the electron emits and reabsorbs a virtual photon while traveling from X to Y.

$$\text{Amp.} = Y 4\pi e^2 \int \frac{1}{\not{p} - m} \gamma_\mu \frac{1}{\not{p} - \not{k} - m} \gamma_\mu \frac{1}{k^2} \frac{1}{\not{p} - m} \times \frac{d^4k}{(2\pi)^4} X$$

$$= Y \frac{1}{\not{p} - m} C \frac{1}{\not{p} - m} X$$

where

$$C = 4\pi e^2 \int \gamma_\mu \frac{1}{\not{p} - \not{k} - m} \gamma_\mu \frac{1}{k^2} \frac{d^4k}{(2\pi)^4}$$

C is an invariant of the form $C = A(p^2)\not{p} + B(p^2)$. What is its physical meaning? Suppose C is small. Then we can write the first two terms as

$$Y \frac{1}{\not{p} - m} X + Y \frac{1}{\not{p} - m} C \frac{1}{\not{p} - m} X = Y \frac{1}{\not{p} - m - C} X$$

This is in virtue of

$$\frac{1}{\not{p} - m - C} = \frac{1}{\not{p} - m} + \frac{1}{\not{p} - m} C \frac{1}{\not{p} - m} + \frac{1}{\not{p} - m} C \frac{1}{\not{p} - m} C \frac{1}{\not{p} - m} + \cdots$$

(a special case of a general relation for any two operators A, B,

$$\frac{1}{A - B} = \frac{1}{A} + \frac{1}{A} B \frac{1}{A} + \frac{1}{A} B \frac{1}{A} B \frac{1}{A} + \cdots \Big)$$

If C were just a number we could consider it as a correction to the mass.

The first and second terms in this series are the amplitudes of the electron propagator with zero and with one virtual photon, respectively. It is easy to verify that the third term is a contribution due to two photons,

$$\text{Amp.} = Y \frac{1}{\not{p} - m} C \frac{1}{\not{p} - m} C \frac{1}{\not{p} - m} X$$

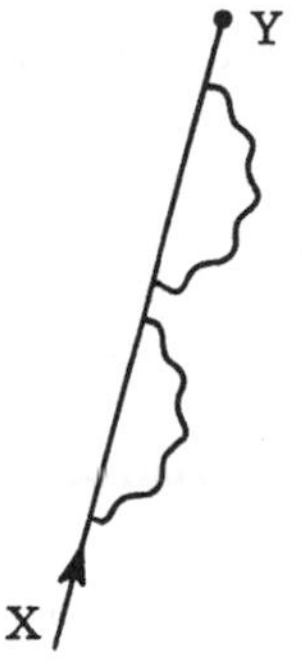

the fourth term, to three photons, etc. However, these diagrams include only processes in which there is only one photon at any given time. For instance, two other ways in which two virtual photons appear are given in Fig. 28-1.

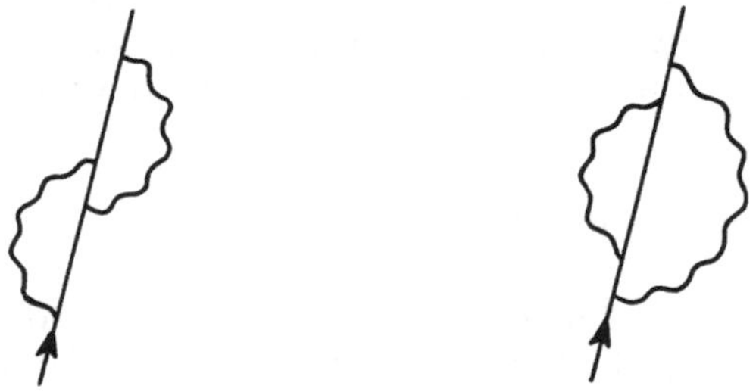

FIG. 28-1

We shall for the moment forget these diagrams; they contribute terms of order e^4 to C when we write the total amplitude for propagation of an electron between X and Y in the form

$$\frac{1}{\not{p} - m - C} = \frac{1}{\not{p} - m - A\not{p} - B}$$

where A and B are functions of p^2. The pole of this propagator gives the relation between energy and momentum for the free particle, and therefore the experimentally observed mass m_{exp}.

By rationalizing,

$$\frac{1}{(1 - A)\not{p} - (m + B)} = \frac{(1 - A)\not{p} + (m + B)}{(1 - A)^2 p^2 - (m + B)^2}$$

we see that the pole is the solution of

$$[1 - A(p^2)]^2 p^2 - [m + B(p^2)]^2 = 0$$

Incidentally, if there happens to be more than one pole, this could be interpreted as another particle (maybe the μ meson). Assuming $A \ll 1$ and $B \ll m$ we can set $A(p^2) = A(m^2)$ and $B(p^2) = B(m^2)$. Then

$$p^2 = \left[\frac{m + B(m^2)}{1 - A(m^2)}\right]^2 = m_{exp}^2$$

or

$$m_{exp} = (m + B)/(1 - A) = m + \delta m$$

where

$$\delta m = B(m^2) + mA(m^2)$$

Thus the propagator has a pole at $\not{p} = m_{exp}$, so for p^2 near m_{exp}^2 it behaves as some constant (the residue at the pole) times $(\not{p} - m_{exp})^{-1}$. We write the residue at $p^2 = m_{exp}^2$ as $1 + r$. We can rewrite the propagator in the form

$$(1 + r)/(\not{p} - m_{exp})$$

[r can be expressed in terms of A, B, and their derivations $A'(p^2)$, $B'(p^2)$ evaluated at $p^2 = m_{exp}^2$]. The change from the usual form $(\not{p} - m_{exp})$ can be interpreted as a correction in the photon coupling strength [11] [for a term $(1 + r)$ would occur in each propagator if we said the coupling strength is $(1 + r)^{1/2}$ for each coupling with a photon]. The next step, of course, is to evaluate the functions A and B. For this purpose we need to evaluate the integral

$$\int \gamma_\mu \frac{\not{p} - \not{k} + m}{p^2 - 2p\cdot k + k^2 - m^2} \gamma_\mu \frac{d^4 k}{k^2}$$

If we use the relations

$$\gamma_\mu \gamma_\mu = 4$$

$$\gamma_\mu \not{a} \gamma_\mu = -2\not{a}$$

we get rid of the γ_μ. For δm we should set $p^2 = m^2$. We get

$$\int \frac{-2(\not{p} - \not{k}) + 4m}{-2p \cdot k + k^2} \frac{d^4k}{k^2}$$

This integral diverges. Let us look at its value for large k where the first denominator can be approximated by k^2. The term containing $\not{k}$ would then cancel by symmetry. What is left in the integrand is proportional to $k^3\,dk/k^4$ for large k and therefore the integral is logarithmically divergent. Quantum electrodynamics has fallen flat on its face!

Bethe noticed that this is the only infinity that exists in electrodynamics (except another one which we shall discuss later on). Suppose we had a rule so that temporarily the integral was made convergent. For instance, we could assume that the propagator $1/k^2$ must always be multiplied by a relativistically invariant convergence factor $C(k^2)$.

If we let

$$C(k^2) = -[\lambda^2/(k^2 - \lambda^2)]$$

(this is chosen to be 1 for small k^2, but to cut the integral off for very large k^2) the integral can be evaluated. One finds (for methods see reference 11)

$$\delta m = m(3e^2/2\pi)[3 \ln (\lambda/m) + (3/4)]$$

neglecting terms of order m/λ.

If you calculate any process to higher order you will find a term proportional to $\ln(\lambda/m)$ (no problem, for spin-1/2 electrons interacting only with photons yield nothing worse than logarithmic divergences). Now wherever you find m substitute $m_{exp} - \delta m$ and expand to first order in δm. Then the miracle is that the coefficient of $\ln(\lambda/m)$ becomes identically 0. The remaining terms have a definite limit when $\lambda \to \infty$. In other words, the magnitude of the cutoff parameter λ does not appear if we always express the solution of a problem in terms of the experimental mass and let $\lambda \to \infty$, keeping m_{exp} fixed.

Using similar ideas, Bethe tried to calculate the displacement of the energy levels in the hydrogen atom due to the self-energy of the bound electron. This had been prompted by the experiment of Rutherford and Lamb who observed, using microwave techniques, a separation of about 1000 Mc between the $2S_{1/2}$ and $2P_{1/2}$ levels in hydrogen. If we neglected the interaction with the radiation field these levels would be completely degenerate. Bethe made an incomplete calculation, using a nonrelativistic approximation. The rapid development of quantum electrodynamics in 1948-1949 resulted from attempts to formulate his, and Weisskopf's, ideas in a relativistically invariant way and to complete his calculation.

We have found another rule that must be included in quantum electrodynamics: (1) Put in an arbitrary cutoff factor $C(k^2) = -[\lambda^2/(k^2 - \lambda^2)]$ for each propagator $1/k^2$. (2) Express everything in $m_{exp} = m - \delta m$. (3) Take the limit as $\lambda \to \infty$ and keep m_{exp} fixed.

Schwinger subtracted infinities at the integrand stage, but this is extremely difficult to do in practice. It turns out that his method is completely equivalent to the above-mentioned rule.

Is there anything wrong with this procedure? It is just a dirty-looking prescription. Weisskopf once remarked that, only if God had given us a charged and uncharged electron, would we be forced to compute δm.

Actually we have in nature examples for which the cutoff technique does not work, for instance, the π^+, π°, and π^-. The π^+ and π° differ in mass but the calculation diverges quadratically. When we make this calculation we treat them as point particles. Actually, we should include a cloud of nucleon pairs and some people believe that this could cancel the infinities. However, that has never been proved.

29 Quantum Electrodynamics

Is it possible that we can do quantum electrodynamics replacing the propagator $1/k^2$ by

$$-\frac{1}{k^2}\frac{\lambda^2}{(k^2-\lambda^2)}$$

and keep λ finite? Then there would be no divergences and the cutoff λ could be introduced as a new constant. Unfortunately such a theory is not internally consistent. For example, suppose we have an atom in an excited state. We now calculate two probabilities: (1) the probability that it decays (i.e., that it radiates a photon); (2) the probability that it remains in the excited state. The sum of these probabilities differs from unity by a factor proportional to m^2/λ^2. Probability is not conserved! You can also see this if you write the corrected propagator in the form $(1/k^2) - [1/(k^2-\lambda^2)]$. This amounts to introducing a propagator, $-[1/(k^2-\lambda^2)]$, for an extra "photon" or vector particle of mass λ. The minus sign means that it is coupled with $-e^2$ instead of $+e^2$, that is, the coupling of one such photon would have to be with an imaginary coupling constant ie. The Hamiltonian is not Hermitian, so probability is not conserved, and chaos ensues.

Nobody has been able to solve this problem: Find a theory that is consistent with the general principles of quantum mechanics (superposition of amplitudes) and relativity and contains an arbitrary function. You cannot modify the propagator $1/k^2$ without the whole theory collapsing. Note that this difficulty does not occur in the nonrelativistic quantum mechanics, where an arbitrary function, the potential V(r), can be varied over a considerable range. Relativity plus quantum mechanics seems to be exceedingly restrictive, but we are also undoubtedly adding unknown tacit assumptions (such as indefinitely short distances in space).

We have computed the contributions to the self-energy by the sum of the diagrams of Fig. 29-1 and found that it diverges logarithmically. However, we have left out diagrams of the type shown in Fig. 29-2. This term gives a contribution of order e^4 to C, and thus to δm. It varies as $e^4[\ln(\lambda/m)]^2$. It

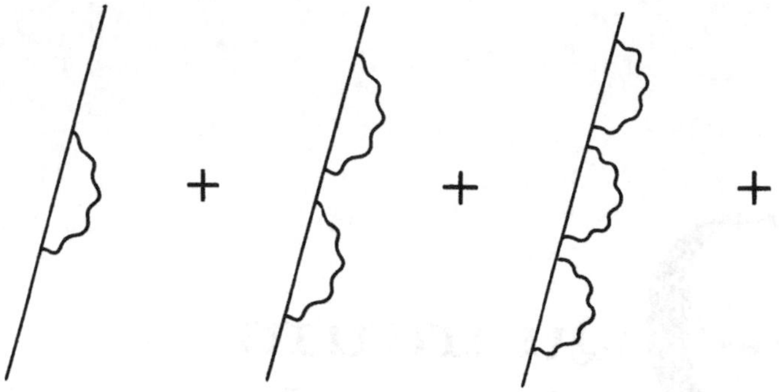

FIG. 29-1

might be possible that if we included all such diagrams, the self-energy would turn out to be finite. Gell-Mann and Low have been able to sum all those terms which are of the highest order in $\ln(\lambda/m)$ and showed that the result is still divergent. It appears that C varies as

$$(\lambda^2/p^2)\,ae^2 + be^2 + \cdots$$

where a, b, ... are numbers.

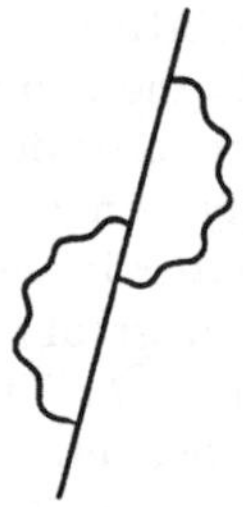

FIG. 29-2

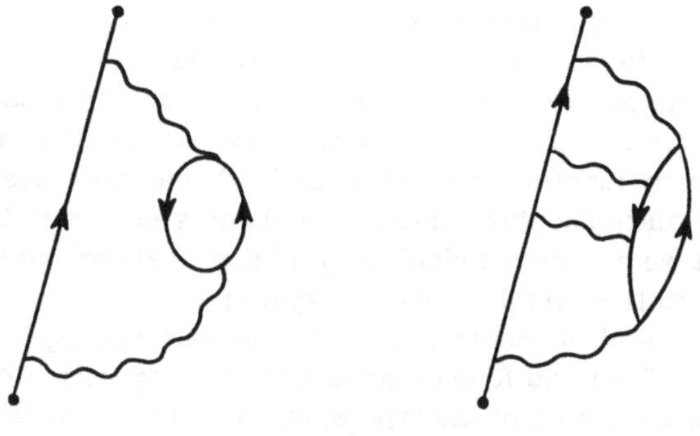

FIG. 29-3

There remains one new type of diagram to be discussed. Namely when a pair produced annihilates itself. As examples, we have the diagrams of Fig. 29-3.

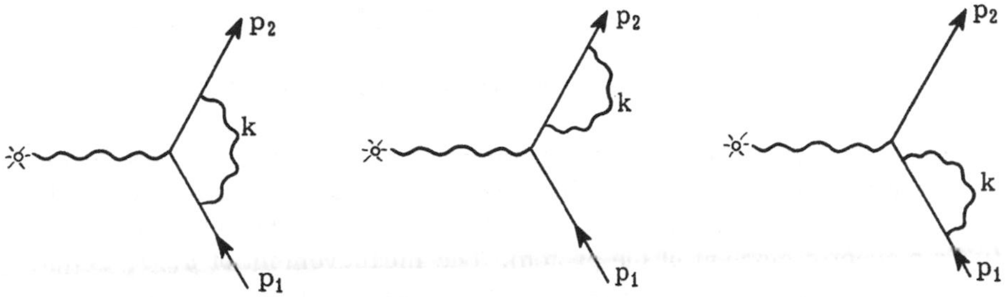

FIG. 29-4

Let us see some of the effects of virtual photons (radiative corrections). Consider, for instance, the scattering of an electron from a potential V(r) (see Lecture 30 for a discussion on the meaning of a potential). The diagrams including virtual photons in lowest order are those of Fig. 29-4. The correction to the coupling strength (factor 1 + r in the propagator) cancels out when we add the contributions of these three diagrams. The net effect, for sufficiently low energies, is to smear out the potential over a Compton wavelength: Very roughly,

$$V(r) \rightarrow V(r) + \text{const}\,(e^2/m^2)\nabla^2 V(r)$$

In an atom this change in the potential will modify the energy levels. Consider the case of hydrogen. Assuming a pure Coulomb potential between the electron and the proton, the Dirac theory predicts that the $2S_{1/2}$ and the $2P_{1/2}$ states have exactly the same energy. However, we have seen that the effective proton-electron potential has also a term proportional to $\nabla^2 V(r) = -4\pi\rho$ (ρ is the charge density of the proton). Since ρ vanishes except at the origin, this affects only the S-state energy, which is shifted by ~1000 Mc. If one includes also the correction due to the diagram of Fig. 29-5 (vacuum polarization), the theory predicts a shift of 1057.3 ± 0.1 Mc. There is a

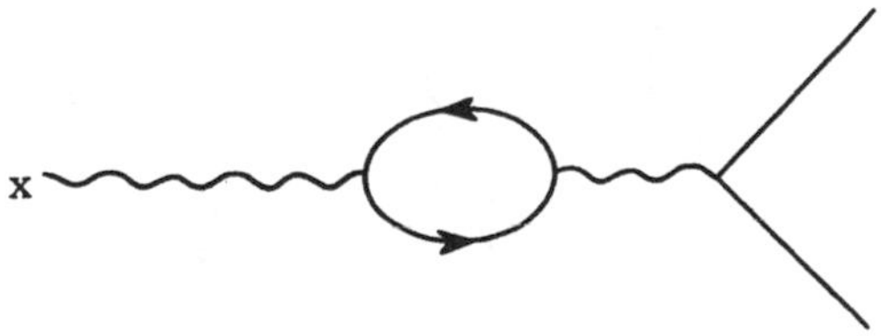

FIG. 29-5

slight disagreement with experiment and it may be necessary to calculate the next order.

If the external potential is a magnetic field then the effect of the virtual photons is to modify the magnetic moment of the electron. Its effective magnetic moment μ_e has been calculated to order e^4 and the result is

$$\mu_e = \mu_0 [1 + (e^2/2\pi) - 0.328 (e^4/\pi^2)] = \mu_0 (1.0011596)\mu_0$$

where $\mu_0 = e/2m$. (The correct coefficient of e^4 was obtained only recently by Petermann and by Sommerfield. The first calculation by Karplus and Kroll gave 2.973.) The magnetic moment is measured by determining μ_e/μ_p (μ_p = magnetic moment of the proton). The measurement of μ_e/μ_p is quite accurate. However there are two conflicting experimental determinations of μ_p/μ_0: one gives $\mu_e/\mu_0 = 1.00146 \pm 0.000012$, and the other gives $\mu_e/\mu_0 = 1.001165 \pm 0.00011$ (reference 12).

Charge Renormalization. I said before that there is another infinity in quantum electrodynamics. We have diagrams of the type given in Fig. 29-6

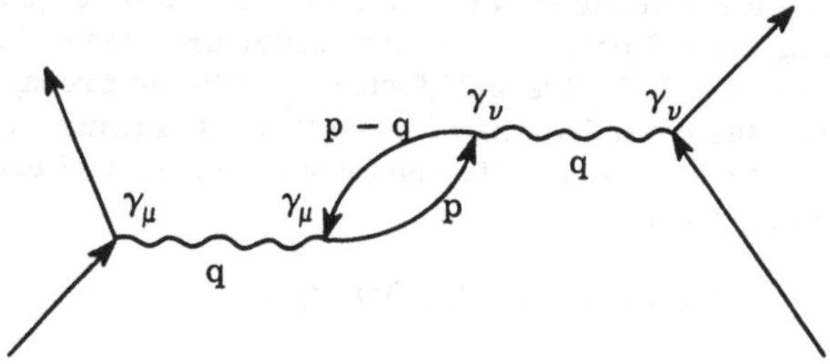

FIG. 29-6

in which there appear virtual electron-positron pairs. We can again sum the diagrams of Fig. 29-7. The corresponding series is

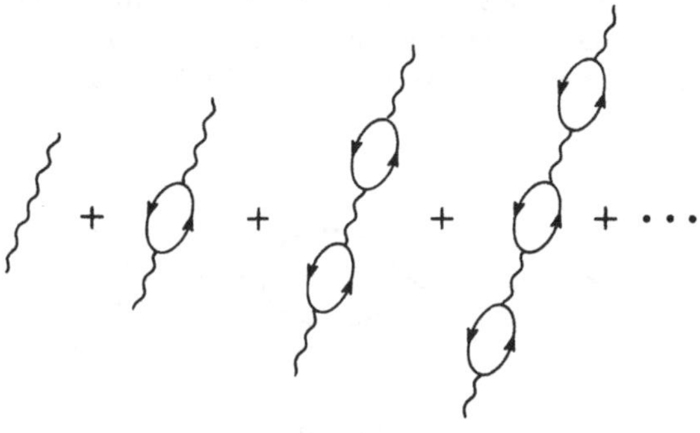

FIG. 29-7

$$\frac{1}{q^2} + \frac{1}{q^2} X \frac{1}{q^2} + \frac{1}{q^2} X \frac{1}{q^2} X \frac{1}{q^2} + \cdots = \frac{1}{q^2 - X}$$

where X is the contribution of the electron-positron loop.

It turns out that for small q^2, $X = q^2 Y$ where Y approaches a constant (this is true in fact to all orders in e^2). Therefore,

$$1/(q^2 - X) = 1/[q^2(1 - Y)]$$

The pole of the propagator is still at $q^2 = 0$. This shows that the rest mass of the photon stays equal to zero. However, the factor $1/(1 - Y)$ will always appear multiplying e^2. Therefore the charge e_{exp} measured experimentally is

$$e_{exp} = e/(1 - Y)^{1/2}$$

This effect is called renormalization of charge. When you calculate Y you again get infinity. But you can correct this logarithmic divergence in the same way as with the mass. Now, we can see a physical example of renormalization of mass: particle with and without charge (see Lecture 28). But we can see as yet no way that the degree of renormalization of charge can have a physical meaning. However, we know

$$e^2_{exp} = 1/137.0369$$

Suppose a future theory predicts some simple result for the theoretical mass. Say the root of a Bessel function, or something similar

$$e^2_{th} = 1/141$$

But the agreement with experiment only results after the charge renormalization correction, and you get $e^2_{exp} = 1/137$. But this is completely speculative!

Now let us see what X is. We have to calculate the contribution of the diagram of Fig. 29-8. Following the electron line along the closed loop we have $(4\pi)^{1/2} e\gamma_\mu$ = amplitude for annihilating the photon, $1/(\not{p} - m)$ = amplitude

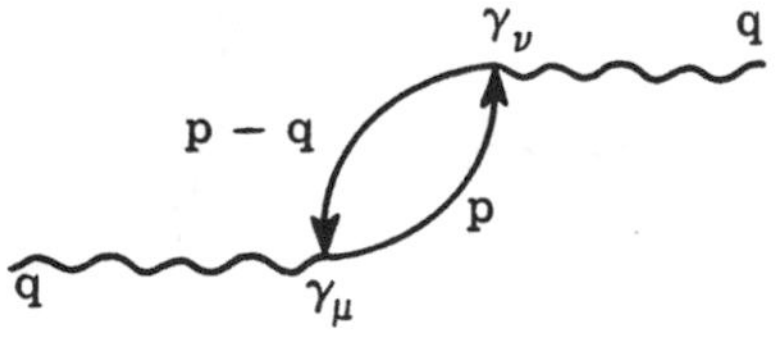

FIG. 29-8

for the propagation of the electron between the two photon vertices, $(4\pi)^{1/2}e\gamma_\nu$ = amplitude for emission of a photon with polarization ν, and $1/(\not{p} - \not{q} - m)$ = amplitude for the propagation of the electron back to its starting point. The total amplitude is therefore

$$4\pi e^2 U_i [1/(\not{p} - \not{q} - m)]\, \gamma_\nu\, [1/(\not{p} - m)]\, \gamma_\mu\, U_i$$

where U_i is the initial state of the electron (which need not satisfy the Dirac equation, since it represents a virtual electron). However, all possible momenta p and initial states U_i can occur. Therefore

$$X_{\nu\mu} = 4\pi e^2 \int \text{spur}\, \{[1/(\not{p} - \not{q} - m)]\, \gamma_\nu\, [1/(\not{p} - m)]\, \gamma_\mu\}\, [d^4 p/(2\pi)^4]$$

The details of how to do the integral are found in reference 11. The suggestion on how to get rid of the infinity was first given by Pauli and Bethe. It is not possible to modify the electron propagator by convergence factor since the resulting expression is not gauge-invariant. Instead, one should calculate using the mass m of the electron and subtract the same integrand with a different mass M. The result is logarithmically divergent but can be taken care of by changing e to e_{exp}.

The renormalization of charge is not only due to virtual electron-positron pairs, but also to every charged particle-antiparticle pair. Is then the renormalized charge of the electron different, say, from that of protons? The answer is no. If the electron-photon coupling is modified by diagrams of the type shown in Fig. 29-9, then the proton-photon coupling has the same diagrams (Fig. 29-10). (Actually there are additional diagrams if we consider

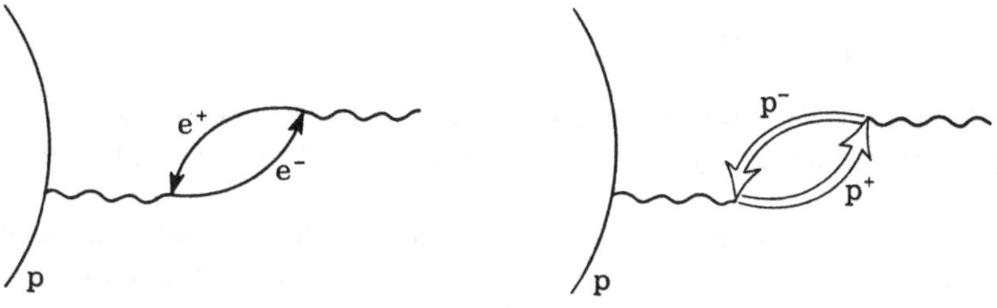

FIG. 29-9

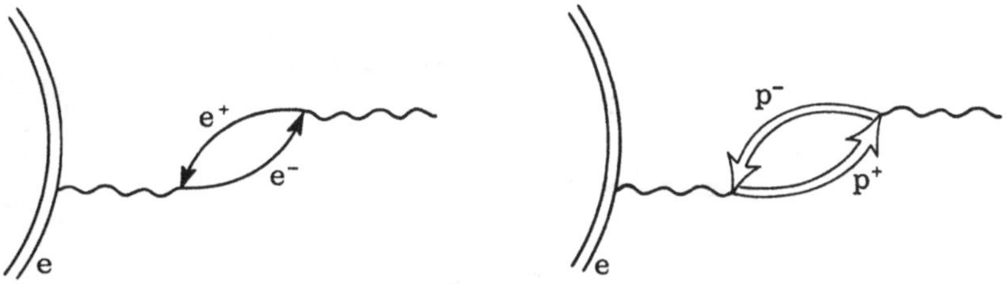

FIG. 29-10

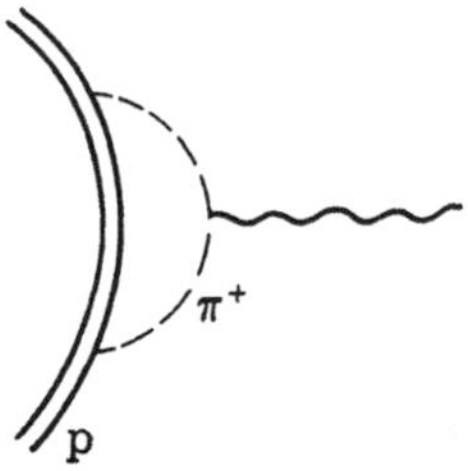

FIG. 29-11

meson theory, Fig. 29-11. They make things like the anomalous moment but do not alter the total charge of the nucleons.)

30 Meson Theory

You are used to seeing the Dirac equation (or similarly the Schrödinger equation) written in the form

$$(i\nabla\!\!\!/ - \mathbf{A} - m)\Psi = 0$$

where **A** is the *external* potential. It is essential to recognize that this result is an approximation to the rules we have given, and can be deduced from them. Therefore we ask: When can part of an interaction be characterized as external potential?

Consider the interaction of an electron with some unspecified machinery that produces virtual photons, the amplitude for producing a virtual photon of momentum q, polarization μ being $A_\mu(q)$. Then the matrix describing the part of the interaction shown is (Fig. 30-1):

$$\int [1/(p\!\!\!/ + q\!\!\!/ - m)]\, \gamma_\mu [1/(p\!\!\!/ - m)]\, A_\mu(q)\, [d^4q/(2\pi)^4]$$

$p + q$ is the "actual" momentum of the electron after absorption of the photon.

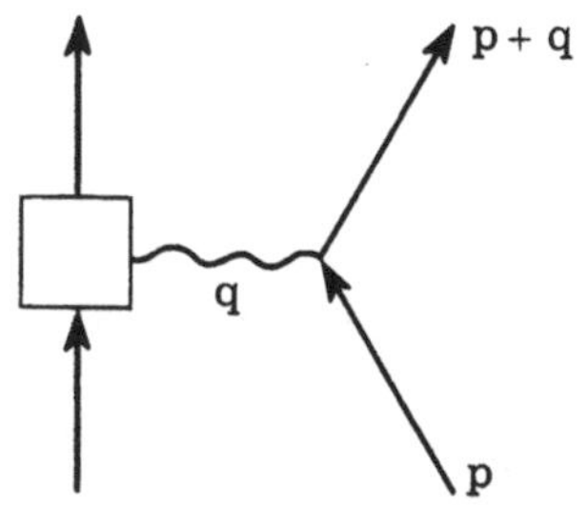

FIG. 30-1

Now the source may emit two, three, or four photons. For two photons the appropriate propagator is

$$\iint \frac{1}{\not p + \not q_1 + \not q_2 - m} \gamma_\nu \frac{1}{\not p + \not q_1 - m} \gamma_\mu \frac{1}{\not p - m} f_{\mu\nu}(q_1,q_2) \frac{d^4 q_1}{(2\pi)^4} \frac{d^4 q_2}{(2\pi)^4}$$

$f_{\mu\nu}(q_1,q_2)$ is the amplitude for emission of two photons, etc.

The important point is that there exist sources that are essentially unaffected by the emission of the first photon. This implies that

$$f_{\mu\nu}(q_1,q_2) = A_\mu (q_1) A_\nu(q_2)$$

That is, for such sources the amplitude to emit a second photon is independent of whether a first photon is emitted or not.† Only then do we say that the source has produced an external potential. [Strictly we assume the amplitude for three photons to be $A_\mu (q_1)A_\nu(q_2)A(q_3)$, etc., for any number.] The total propagator for an electron in an external field is then

$$\frac{1}{\not p - m} + \frac{1}{\not p + \not q - m} \not A(q) \frac{1}{\not p - m} + \frac{1}{\not p + \not q_1 + \not q_2 - m} A(q_2) \frac{1}{\not p + \not q - m} A(q_1)$$

$$\times \frac{1}{\not p - m} + \cdots$$

plus similar terms (integrations understood), since any number of photons can be absorbed.

This series can be summed. We shall show two ways. First, let the "actual" momentum $\not p + \not q_1 + \cdots = \not P$, considered as an operator changing its value after operations with $\not a$ where

$$\not a \equiv \not A(q)[d^4 q/(2\pi)^4]$$

Then the above series can be written

---

†It is evident that one of the requirements is that the source remain in the same state after the emission of one photon. For example, for a heavy particle, recoil upon emission of a photon is negligible.

As another example, which is more subtle, consider a big magnet, containing lots of electrons. Now a single electron does not produce the usual external magnetic field, since the spin is flipped $\sim 1/2$ of the time when a virtual photon is emitted. But for iron the situation is different. Define x and y (for a single electron) by: x = amp. to stay in the same state after emitting one photon ($x^2 \sim 1/2$), y = amp. to change to another state. Let N be the number of electrons in the chunk of iron. Then we have: amp. for an electron to stay in the same state = Nx, amp. for an electron to change state = y, since if the spin flipped we could distinguish the guilty electron. The corresponding probabilities are $(xN)^2$, Ny^2, so that the amplitude to remain unchanged effectively is $O(\sqrt{N})$ greater than to change.

$$\frac{1}{\not p - m} + \frac{1}{\not p - m} \not a \frac{1}{\not p - m} + \cdots$$

which we recognize as the expansion of $1/(\not p - m - \not a)$.

Another way is as follows: Consider the diagrams of Fig. 30-2. There is some "last" photon in each diagram. What is the amplitude Ψ to arrive at

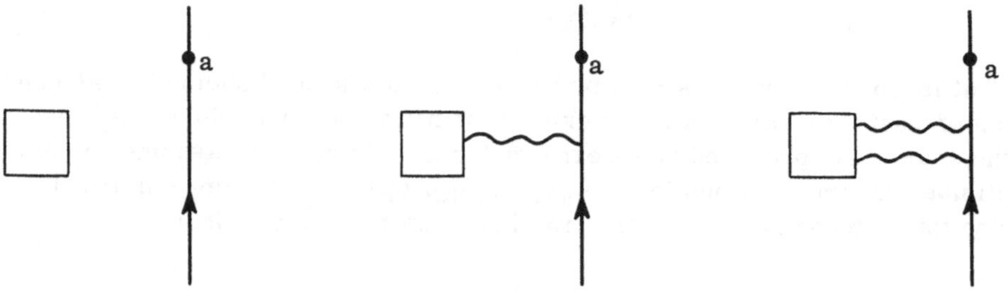

FIG. 30-2

the position a after the last photon is absorbed? Here Ψ is also the amplitude after all but a finite number of photons have been absorbed. Therefore $[1/(\not p - m)] \not a \Psi$ [the amplitude Ψ previous to "last" photon, times the amplitude $\not a$ to scatter "last" photon and propagate $1/(\not p - m)$] plus the amplitude φ that there is no photon is Ψ again:

$$\Psi = [1/(\not p - m)] \not a \Psi + \varphi$$

where

$$(\not p - m)\varphi = 0$$

Therefore,

$$(\not p - m - \not a)\Psi = 0$$

Alternatively, if φ_n amplitude after absorbing n photons,

$$\varphi_{n+1} = [1/(\not p - m)] \not a \varphi_n$$

and

$$\Psi = \sum_{n=0}^{\infty} \varphi_n = \varphi_0 + [1/(\not p - m)] \not a \Psi$$

Meson Theory. At present there is no quantitative meson theory. Present theories are based on an analogy with electrodynamics (Table 30-1). This

Table 30-1

	Meson theory		Electrodynamics	
Propagator:	nucleon	$1/(\not{p} - m_N)$	electron	$1/(\not{p} - m_e)$
	pion (spin 0)	$1/(q^2 - m_\pi^2)$	photon (spin 1)	$-1/q^2$
Coupling:	$4\pi g \bar{\Psi}\gamma_5 \varphi_\pi \Psi_N$		$4\pi e \bar{\Psi}_e \gamma_\mu a_\mu \Psi_e$	

γ_5 appears since the pion is pseudoscalar

theory is clearly a creation of the human mind! It is wrong because nature is more inventive.

If you want, you can write down diagrams, putting in the right bookkeeping to account for charge conservation. But the diagrams correspond to a perturbation expansion, and since $g^2 \cong 15$ (not 1/137) each successive term is more important than the one before!

The γ_5 coupling is called the pseudoscalar coupling (PS). Also possible is pseudovector (PV)($\gamma_5 \not{q}$). However there is a prejudice against PV coupling—it cannot be renormalized, each successive order diverges worse than the preceding one because of the extra momenta in the numerator.

It seems that if the theory were basically correct, experiment would have already given us some hints as to what the correct approximations are.

31 Theory of β Decay

We have discussed electrodynamics. The only other process about which quantitative calculations can be made is β decay. It was first observed in some nuclei that $N \to P + e + \bar{\nu}$ ($\bar{\nu}$ is defined to be the antineutrino). You know about the neutrino. Its existence was postulated in order to preserve conservation of energy, momentum, and spin. It can have 0 mass and has spin 1/2.

Around 1934 Fermi proposed that the amplitude for this transition be written in the form $g(\Psi_N \Psi_P \Psi_e \Psi_\nu)$ where Ψ is the wave function of the corresponding particle. In the absence of any gradient of Ψ, the electron energy spectrum could be obtained from the density of final states alone. (This at first appeared to be incorrect experimentally. So Konopinski and Uhlenbeck suggested that one should include gradients in the wave functions, and obtained a spectrum in agreement with the experiments. But all the experiments were wrong because back scattering in the foil had not been taken into account. Miss Wu did the experiment with thinner and thinner foils and found this out. Fermi had been right. End of Konopinski and Uhlenbeck.)

One question is apparent: Each wave function has four components, so which component do you put in the coupling? There is a total of 256 possibilities. Physically what we are asking is how does it depend on the spin of the particles? What we have to find are the couplings that are invariant under rotations and Lorentz transformations.

One possibility is $C_S(\bar{\Psi}_P \Psi_N)(\bar{\Psi}_e \Psi_\nu)$, known as scalar coupling. Another possibility is vector coupling,

$$C_V(\bar{\Psi}_P \gamma_\mu \Psi_N)(\bar{\Psi}_e \gamma_\mu \Psi_\nu)$$

which is also invariant. (This is the one Fermi proposed as an example.)

You can keep inventing things. Using the antisymmetric tensor of the second rank

$$\sigma_{\mu\nu} = (i/2)(\gamma_\mu \gamma_\nu - \gamma_\nu \gamma_\mu)$$

we can construct the tensor coupling

$$C_T(\bar{\Psi}_p \sigma_{\mu\nu} \Psi_N)(\bar{\Psi}_e \sigma_{\mu\nu} \Psi_\nu)$$

The remaining ones are the axial vector coupling

$$C_A(\bar{\Psi}_p \gamma_\mu \gamma_5 \Psi_N)(\bar{\Psi}_e \gamma_\mu \gamma_5 \Psi_\nu)$$

and the pseudoscalar coupling

$$C_P(\bar{\Psi}_p \gamma_5 \Psi_N)(\bar{\Psi}_e \gamma_5 \Psi_\nu)$$

The correct coupling could be any linear combination of these. We have assumed invariance under reflections (conservation of parity); $C_S'(\bar{\Psi}_p \gamma_5 \Psi_N) \times (\bar{\Psi}_e \Psi_\nu)$ is invariant under rotation and Lorentz transformation, but changes sign under reflections. The same applies for the other four couplings if we insert an extra γ_5. If we combine couplings that change sign under reflections with couplings that do not change sign, parity would be violated. Consequently these couplings were ignored until the difficulties of the K^+-meson decay ($\tau - \theta$ puzzle). The K^+ decays into 2π's and 3π's, but the parity of the final state is different in the two cases. Lee and Yang proposed several experiments to find out whether this apparent lack of conservation of parity was characteristic of the weak decays. According to Wu's experiment on Co^{60} (see Lecture 7) the electrons came out backward with respect to the orientation of the spin. This means that you can associate a rotation with a direction in space, so reflection symmetry is not obeyed. Mathematically the violation of conservation of parity in β decay meant that the couplings which change sign under reflections had to be included, in addition to the ordinary ones. If the C's are real, the theory is invariant under time reversal. However, immediately after the failure of parity was demonstrated, people also began to doubt the validity of time-reversal invariance. So there were ten complex C's or twenty constants in the theory.

The next proposal was made by Lee and Yang and also independently by Landau and by Salam. Their idea was that the lack of conservation of parity was due to the neutrino, which must always spin to the left. (Originally they had it spinning to the right, which is wrong.) Recall when we discussed relativistic particles of spin 1/2 that the simplest representation was a two-component amplitude satisfying equations

$$(E - \sigma \cdot \mathbf{P})u = 0$$

or

$$(E + \sigma \cdot \mathbf{P})v = 0$$

What Lee and Yang said is that the neutrino could only exist in one of these states. The equation for the neutrino turns out to be

$$(E + \sigma \cdot \mathbf{P})v = 0$$

Recall that for an electron we had to write

$$(E - \sigma \cdot \mathbf{P})u = mv$$

$$(E + \sigma \cdot \mathbf{P})v = mu$$

if we insisted in a first-order equation. However it is clear that v also satisfies the second-order equation

$$(E - \sigma \cdot \mathbf{P})(E + \sigma \cdot \mathbf{P})v = m^2 v$$

Gell-Mann and I proposed that the electron was also represented by a two-component spinor v. Then the β-decay couplings consist only of two-component wave functions v. The only relativistic invariant combination using no gradients is

$$G(v_p^* \sigma_\mu v_N)(v_e^* \sigma_\mu v_\nu) \qquad \sigma_4 = 1,\ \sigma_{1,2,3} = \text{Pauli spin matrices}$$

The same proposal was also made, possibly somewhat earlier, by Marshak and Sudarshan.

We have now a unique theory for the β decay with only one coupling constant, G. When it was proposed it disagreed with at least three accepted experiments, but all of them have been found to be wrong.

I was tempted to teach quantum electrodynamics with a two-component wave function. The only difficulty is that you could not read any of the literature. For this reason we shall also write the β coupling in the four-component representation. In our representation for the γ matrices

$$i\gamma_5 = \begin{pmatrix} -1 & 0 & 0 & 0 \\ 0 & -1 & 0 & 0 \\ 0 & 0 & 1 & 0 \\ 0 & 0 & 0 & 1 \end{pmatrix}$$

Let

$$a = (1/2)(1 + i\gamma_5) = \begin{pmatrix} 0 & 0 & 0 & 0 \\ 0 & 0 & 0 & 0 \\ 0 & 0 & 1 & 0 \\ 0 & 0 & 0 & 1 \end{pmatrix} = \begin{pmatrix} 0 & 0 \\ 0 & 1 \end{pmatrix}$$

Then if

$$\Psi = \begin{pmatrix} u \\ v \end{pmatrix}$$

where u and v are two-component wave functions

$$a\Psi = \begin{pmatrix} 0 \\ v \end{pmatrix}$$

In the same way,

$$\bar{a} = (1 - i\gamma_5)/2 = \begin{pmatrix} 1 & 0 \\ 0 & 0 \end{pmatrix}$$

and

$$\bar{a}\Psi = \begin{pmatrix} u \\ 0 \end{pmatrix}$$

Here a and $\bar{a}$ are projection operators. You can verify that

$$a^2 = a \qquad \bar{a}^2 = \bar{a} \qquad a\bar{a} = \bar{a}a = 0 \qquad \bar{a} + a = 1$$

and a projects out the v component of Ψ. Therefore in four-components the coupling becomes

$$G(\overline{a\Psi}_P \gamma_\mu a\Psi_N)(\overline{a\Psi}_e \gamma_\mu a\Psi_\nu)$$

Since $\bar{a}\gamma_\mu = \gamma_\mu a$ and aa = a, this can be simplified to

$$G(\Psi_P \gamma_\mu a\Psi_N)(\bar{\Psi}_e \gamma_\mu a\bar{\Psi}_\nu)$$

After 23 years we come back to Fermi!

Fermi's rule is just modified by replacing $a\Psi$ for every Ψ. It took 23 years to find the a. It is easy to verify that if one applies this substitution to all the β couplings, then the scalar, tensor, and pseudoscalar components vanish, and the vector and the axial vector give the above result. Historically, Salam, and Landau, and Lee and Yang, proposed that the neutrino wave function be always multiplied by a. Afterward I proposed the same for the electron and muon, but hesitated to apply it to neutron and proton because I believed there were some wrong experiments. Finally Marshak and Sudarshan, and Gell-Mann and I, proposed the general rule, every Ψ replaced by $a\Psi$.

Let us find out now what is the physical content of this theory. For this purpose we look at the decay of a polarized neutron. For simplicity we neglect the motion of the nucleons (let the nucleon mass $M \to \infty$) and the spin of the proton. The amplitude m for this process is

$$m = G(\bar{U}_p \gamma_\mu a U_N)(\bar{U}_e \gamma_\mu a U_\nu)$$

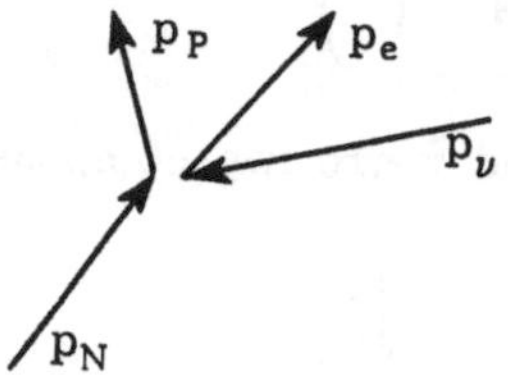

We are interested in m^*m summed over the two-spin states of the proton. By using projection operators $(1/2)(1 + i\not{W}_N \gamma_5)$ and $(1/2)(1 + i\not{W}_e \gamma_5)$ for the spins of the neutron and electron,

$$\sum_{\text{proton spin}} m^*m = G^2 \,\text{spur}\,\{\gamma_\nu a(\not{p}_p + M)\gamma_\mu a(\not{p}_N + M) \times [(1 + i\not{W}_N \gamma_5)/2]\} \times \text{spur}\,\{\gamma_\nu a(\not{p}_e + m)[(1 + i\not{W}_e\gamma_5)/2]\,\gamma_\mu a\not{p}_\nu\}$$

Consider first the spur containing the nucleon coordinates.† Cancel one of the a's using the fact that $a^2 = a$ and $\bar{a}a = 0$. Then

$$\text{spur}\,\{a\gamma_\nu(\not{p}_p + M)\gamma_\mu a(\not{p}_N + M)[(1 + i\not{W}_N\gamma_5)/2]\} = (1/2)\,\text{sp}\,[\gamma_\nu\not{p}_p\gamma_\mu(\not{p}_N + M)(1 + i\not{W}_N\gamma_5)\bar{a}]$$

Now

$$(1 + i\not{W}_N\gamma_5)\bar{a} = (1 - \not{W}_N)\bar{a}$$

Since the spur of an odd number of γ matrices vanishes, we are left with

$$(1/2)\,\text{sp}\,[\gamma_\nu\not{p}_p\gamma_\mu(\not{p}_N - M\not{W}_N)(1 - i\gamma_5)]$$

Take the z axis along the polarization of the neutron. In the limit $M \to \infty$ this becomes

$$(M^2/2)\,\text{sp}\,[\gamma_\nu\gamma_t\gamma_\mu(\gamma_t + \gamma_z)(1 - i\gamma_5)]$$

Using

---

† Notice that for a neutrino the usual normalization $\bar{U}U = 2m$ cannot be applied. However, we are calculating spurs. The only thing that happens in the limit $m \to 0$ is that the projection operator becomes simply $\not{p}$.

$$(1/4)\ \mathrm{sp}\ \not{a}\not{b}\not{c}\not{d} = (a \cdot b)(c \cdot d) - (a \cdot c)(b \cdot d) + (a \cdot d)(b \cdot c)$$

we get

$$(1/4)\ \mathrm{sp}\ \gamma_\nu \gamma_t \gamma_\mu \gamma_t = 2\delta_{\nu t}\delta_{\mu t} - \delta_{\mu\nu}$$

$$(1/4)\ \mathrm{sp}\ \gamma_\nu \gamma_t \gamma_\mu \gamma_z = \delta_{\nu t}\delta_{\mu z} + \delta_{\nu z}\delta_{\mu t}$$

(An easier way is to work out each special case for μ and $\nu = t,x,y,z.$) Also,

$$(1/4)\ \mathrm{sp}\ \gamma_\nu \gamma_t \gamma_\mu \gamma_t \gamma_5 = 0$$

$$(1/4)\ \mathrm{sp}\ \gamma_\nu \gamma_t \gamma_\mu \gamma_z \gamma_5 = -\delta_{\nu x}\delta_{\mu y} + \delta_{\nu y}\delta_{\mu x}$$

Therefore,

$$\begin{aligned}(1/4)\ \mathrm{sp}\ [\gamma_\mu \gamma_t \gamma_\mu (\gamma_t + \gamma_z)(1 - i\gamma_5)] = {} & 2\delta_{\mu t}\delta_{\nu t} - \delta_{\mu\nu} + \delta_{\mu t}\delta_{\nu z} \\ & + \delta_{\mu z}\delta_{\nu t} \\ & -i(\delta_{\mu x}\delta_{\nu y} - \delta_{\mu y}\delta_{\nu x})\end{aligned}$$

The spur containing the electron neutrino can also be simplified into the form

$$(1/2)\ \mathrm{sp}\ [\gamma_\mu \not{p}_\nu \gamma_\nu (\not{p}_e - m\not{W}_e)(1 - i\gamma_5)]$$

Now we need to calculate

$$\mathrm{sp}\ [\gamma_\nu \gamma_t \gamma_\mu (\gamma_t + \gamma_z)(1 - i\gamma_5)]\ \ \mathrm{sp}\ [\gamma_\mu \not{p}_\nu \gamma_\nu (\not{p}_e - M\not{W}_e)(1 - i\gamma_5)]$$

Substituting our expression for the left-hand-side spur we get

$$\begin{aligned}4\ \mathrm{spur}\ [(&2\gamma_t \not{p}_\nu \gamma_t - \gamma_\mu \not{p}_\nu \gamma_\mu + \gamma_t \not{p}_\nu \gamma_z + \gamma_z \not{p}_\nu \gamma_t - i\gamma_x \not{p}_\nu \gamma_y \\ & + i\gamma_y \not{p}_\nu \gamma_x)(\not{p}_e - M\not{W}_e)(1 - i\gamma_5)]\end{aligned}$$

Evaluating the spur this expression reduces to

$$16\,(E_\nu + P_{\nu_z})(E_e - MW_{e_t})$$

In the rest frame of the electron $W_{e_t} = 0$, $\mathbf{W} = \varepsilon\,(\mathbf{P}_e/P_e)$, where $\varepsilon = +1$ for an electron spinning to the right and $\varepsilon = -1$ spinning to the left. Since W_e transforms like a 4-vector we have in the laboratory frame

$$W_{e_t} = \gamma\,(0 + v_e \varepsilon) = \varepsilon(E_e/m)v_e$$

Finally, if we let θ_ν be the angle between the spin of the neutron and the direction of the emitted antineutrino we get

$$\sum m^*m = 4G^2M^2E_eE_\nu(1 + \cos\theta_\nu)(1 - \varepsilon v_e)$$

This says the following:

Prob. for electrons spinning to the left

$$= (1/2)[1 + (v_e/c)] \approx 1 \text{ for } v \approx c$$

Prob. for electrons spinning to the right

$$= (1/2)[1 - (v_e/c)] \approx 0 \text{ for } v \approx c$$

Hence, electrons are polarized spinning to left when coming out of β decay. Neutrinos must always spin to the left (antineutrino to the right). Notice that relativistic electrons behave like neutrinos because their rest mass can be neglected.

While the electron is emitted isotopically the antineutrino is emitted preferentially along the spin of the neutron with a $(1 + \cos\theta)$ distribution.

We can see that the results are in agreement with the $Co^{60} \to Ni^{60}$ experiment. The spins are 5 and 4, so the total angular momentum changes by 1. The neutrino is emitted preferentially along the spin of the Co^{60} nucleus carrying 1/2 units of angular momentum along its direction of propagation. To preserve total angular momentum the electron must therefore be emitted backward (Fig. 31-1).

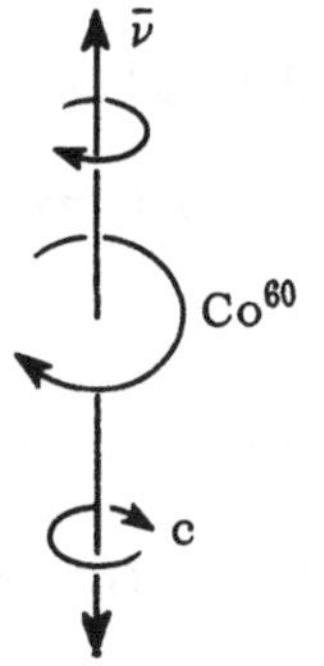

FIG. 31-1

The energy spectrum dN of the electrons depends only on the density of final states (Lecture 16):

$$dN \simeq (E_e - E_0)^2 P_e E_e \, dE_e$$

where

$$E_0 = M_N - M_P$$

The neutron decay rate is

$$1/\tau = [G^2/(2\pi)^3] \int_0^{E_0} (E_e - E_0)^2 P_e E_e \, dE_e$$

32 Properties of the β-Decay Coupling

Consider the process A + C → B + D

$$\text{Amp.} = (\overline{B}A)(\overline{D}C) = (v_B^*\sigma_\mu v_A)(v_D^*\sigma_\mu v_C)$$

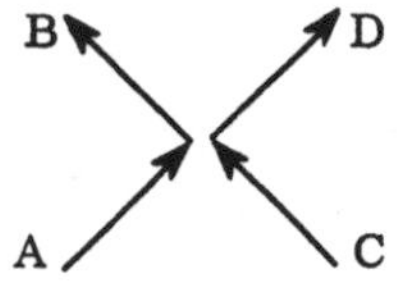

where the v's are two-component wave functions, $\sigma_t = 1$, $\sigma_{x,y,z}$ = Pauli spinors. The neutron decay corresponds to $N \to P + e + \bar{\nu}$ (or $N + \nu \to P + e$). If we expand the coupling in terms of the components of v we get

$$(\overline{B}A)(\overline{D}C) = 2(B_1D_2 - B_2D_1)^* (A_1C_2 - A_2C_1)$$

where

$$v_A = \begin{pmatrix} A_1 \\ A_2 \end{pmatrix}, \text{ etc.}$$

Exchanging A ↔ C or B ↔ D changes only its sign. Therefore,

$$(v_B^* \sigma_\mu v_A)(v_D^*\sigma_\mu v_C) = -(v_D^*\sigma_\mu v_A)(v_B^*\sigma_\mu v_C)$$

Clearly, it is also equal to $2(v_{\overline{C}}^* v_A)(v_D^* v_{\overline{B}})$ where $v_{\overline{B}} = \sigma_y v_B^*$. You can easily verify that $v_{\overline{B}}$ is the wave function for the antiparticle $\overline{B}$. In the four-component representation we have

$$\begin{aligned}(\overline{B}A)(\overline{D}C) &= (\overline{\Psi}_B \gamma_\mu a\Psi_A)(\overline{\Psi}_D\gamma_\mu a\Psi_C) \\ &= (\overline{\Psi}_D\gamma_\mu a\Psi_A)(\overline{\Psi}_B \gamma_\mu a\Psi_C) \\ &= 2(\overline{\Psi}_{\overline{C}} a\Psi_A)(\overline{\Psi}_D\bar{a}\Psi_{\overline{B}}\end{aligned}$$

where

$$a = (1/2)(1 + i\gamma_5)$$

The last expression is the simplest for calculations. Just treat B and C as if they were antiparticles. For instance, the neutron decay amplitude m is $2G(U_{\_\nu}\, aU_N)(U_e\bar{a}\,U_{\_P})$. Then

$$\sum_{\text{proton spin}} m^* m = 4G^2\, sp\{\bar{a}(-\not p_\nu)a(\not p_N + M)[(1 + i\not W_N\gamma_5)/2]\}$$

$$sp\{a(\not p_e + m_e)[(1 + i\not W_e\gamma_5)/2]\,\bar{a}(-\not p_P + M)]$$

$$= (1/4)G^2\, sp[\not p_\nu(\not p_N - M\not W_N)]\, sp[(\not p_e - m_e\not W_e)\not p_P]$$

$$= 4G^2 p_\nu(p_N - MW_N)(p_e - m_eW_e)p_P$$

In the limit $M \to \infty$ this reduces to

$$= 4G^2M^2E_eE_\nu(1 + \cos\theta_\nu)(1 - \varepsilon v_e)$$

Compare with the amount of labor used in calculating this result using

$$m = G(\bar{U}p\gamma_\mu aU_N)(\bar{U}_e\gamma_\mu aU_\nu) \qquad \text{(Lecture 28)}$$

We have assumed, using the convention that the electron is a particle (going forward in time), that the proton and the neutron are also particles [coupling $(\bar{P}N)(\bar{e}\nu)$]. Then the neutrino angular distribution varies as $1 + \cos\theta$ with respect to the neutron spin, while the electrons come out isotropically. If we assume that N and P are antiparticles [coupling $(\bar{N}P)(\bar{e}\nu)$], we find that the neutrino comes out isotropically while the electron varies as $1 + \cos\theta_e$. Telegdi et al.[13] measured the angular distribution of electrons $(1 + A\cos\theta_e)$ and neutrinos $(1 + B\cos\theta_\nu)$ from polarized neutrons and found $A = -0.09 \pm 0.03$, $B = +0.88 \pm 0.15$. This agrees with the usual convention that N and P are particles.

To account for the μ decay we assume also a coupling $(\bar{\nu}\mu)(\bar{e}\nu)$, provided the μ^- is a particle like the electron. Calculating the electron energy spectrum dN we find (neglecting the electron mass compared to the electron momentum and the mass of the muon)

$$dN = 2x^2(3 - 2x)\,dx$$

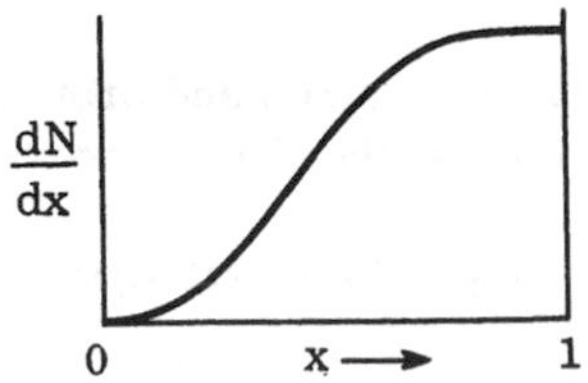

where $x = P_e/P_m$ and $P_m = m_\mu/2$ is the maximum energy of the electron. If the $\bar{\mu}$ is the antiparticle the coupling is, instead,

$$(\bar{\mu}\nu)(\bar{e}\nu)$$

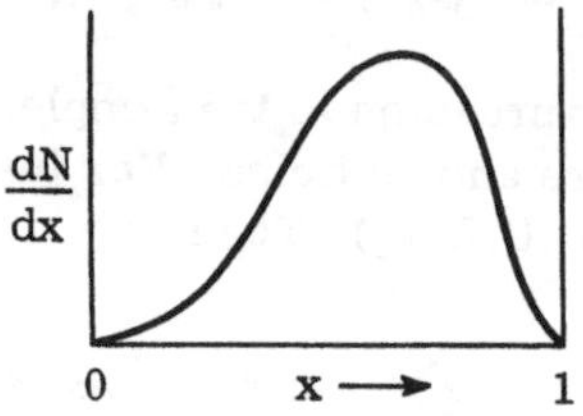

and the electron spectrum turns out to be

$$dN = 12x^2(1 - x)$$

The measurements agree with the spectrum corresponding to $\bar{\mu}$ being a particle. All particles coming from β decay are left-handed.

It turns out that the coupling constant for the neutron and μ decay are the same. If we write it in the form $\sqrt{8}\,G$ (to get the "old" G)

$$GM_p^2 = (1.01 \pm 0.01) \times 10^{-5}$$

(The proton mass has been introduced to make G dimensionless.) We say that for β-decay processes the transition amplitude is proportional to J^*J, where J is a sum of terms of the form

$$J = \Sigma\,(\bar{B}A) \qquad \bar{B}A = (\bar{\Psi}_B \gamma_\mu a \Psi_A)$$

over various combinations of particles A, B. What are these particles? So far we have only been able to establish the part that does not involve strange particles

$$J = (\bar{e}\nu) + (\bar{N}P) + (\bar{\mu}\nu) + (\text{strange particles})$$

The cross term $(\bar{P}N)(\bar{e}\nu)$ gives N decay, $(\bar{\nu}\mu)(\bar{e}\nu)$ gives μ decay, and $(\bar{\nu}\mu)(\bar{N}P)$ gives μ capture. [Note that $(\bar{A}B)^* = (\bar{B}A)$.]

There is only one more thing to say about β decay. Consider again the neutron decay

$$m = \sqrt{8}\,G\{\bar{\Psi}_P[(\gamma_\mu + i\gamma_\mu\gamma_5)/2]\Psi_N\}\,(\bar{\Psi}_e\gamma_\mu a\Psi_\nu)$$

Ordinarily the neutron and proton move very slowly. Therefore it is very useful to look at the N.R. approximation. We have

$$\bar{\Psi}_P\gamma_\mu\Psi_N = (1/2M)\bar{\Psi}_P(\not{p}_P\gamma_\mu + \gamma_\mu\not{p}_N)\Psi_N$$

If the proton and neutron are standing still,

$$\not{p}_P = \not{p}_N = M\gamma_t$$

and

$$\bar{\Psi}_P\gamma_\mu\Psi_N = \bar{\Psi}_P\Psi_N\delta_{\mu 4} = \begin{cases} 2M\delta_{\mu 4} & \text{proton spin parallel to neutron spin} \\ 0 & \text{proton spin not parallel to neutron spin} \end{cases}$$

This is the Fermi part of the coupling.

In a similar manner it follows that

$$\bar{\Psi}_P i\gamma_\mu\gamma_5\Psi_N = \bar{\Psi}_P\sigma_\mu\Psi_N \qquad \mu = 1,2,3$$

$$= 0 \qquad \mu = 4$$

This is the Gamow-Teller contribution. It was soon discovered that the Fermi coupling was not sufficient, since it cannot change the total angular momentum of the nucleus. So Gamow and Teller suggested a term proportional to σ, which can take off unit angular momentum.

We have proposed that the three terms of the current that do not involve strange particles have the same amplitude. However, the π interaction will modify the effective coupling. It is possible to arrange the factors so that the Fermi part is not changed. However, the Gamow-Teller part must be multiplied by a factor $x = 1.25$ (reference 14). The O^{14} decay is a $0 \rightarrow 0$ transition, so the Gamow-Teller contribution vanishes. The coupling constant agrees with that obtained from the lifetime of the μ within 5 per cent.

Recently a striking confirmation of the theory was the observation of the π-e decay. The absolute rate of the π-μ and the π-e decay cannot be calculated. But you can get the ratio. Until this summer the π-e decay had not been identified.

33 Summary of the Course

We have given the rules for processes which involved a small number of particles. Since processes involving large numbers of particles can be understood in terms of the fundamental processes, we have in a sense described all of physics, as follows:

Rules to calculate im		
Propagator:	spin 0	$i/(p^2 - m^2)$
	spin 1/2	$i/(\not{p} - m)$
	photon	$-(i/k^2)$
Electromagnetic coupling:	spin 1/2	$-i(4\pi)^{1/2} e\bar{v}\not{\varepsilon}u$
	spin 0	$-i(4\pi)^{1/2} e(p_1 + p_2)\cdot\mathcal{E} + i4\pi e^2\, \mathcal{E}_1\cdot\mathcal{E}_2$
β-decay coupling:	J*J	

Aside from these rules when the intermediate momenta are not definite, we must integrate over $d^4p/(2\pi)^4$.

For a closed loop take the spur with a minus sign.

If you have identical particles the exchange amplitude gets a + sign for integral spins and a − sign for half-integral spin. Challenge: This last rule is not independent of the other rules. It is necessary in order to get consistent probabilities. If I use a + sign for spin 1/2, I get nonsensical results. However, I do not have a complete proof.

You do not know anything until you have practiced. You are now able to calculate many of the problems of physics by yourself. You still cannot do everything: example—the many-electron atom. The answers are contained in these rules but you will have to learn to use them in their nonrelativistic form, where they correspond to the Schrödinger equation. Also it will not be easy to read papers that are based on the formal theory. However, first learn what the physical problem is, and then try to solve it. Finally, there is one branch of physics that is not contained in these lectures—the Chew-Low theory and dispersion relations. This is an approach to what might be thought to be a field theory of strong couplings. As a result of a study of π-meson nucleon scattering and π-meson photoproduction they have found an approximate formula for the π-nucleon interaction amplitude. It is

$$(f/\mu)\gamma_5 \not{q} \qquad f^2 = 0.08$$

It works only if the energy is not too large.

References

1. P. A. M. Dirac, "Principles of Quantum Mechanics," 4th ed., Oxford, New York, 1958, Chap. 1.
2. R. P. Feynman, Revs. Modern Phys., **20**, 367 (1948).
3. D. Bohm, "Quantum Theory," Prentice-Hall, Englewood Cliffs, N.J., 1951, Chap. 6.
4. W. Pauli, Phys. Rev., **58**, 716 (1940).
5. G. Lüders and B. Zumino, Phys. Rev., **110**, 1450 (1958).
6. W. Pauli and V. Weisskopf, Helv. Phys. Acta, **7**, 709 (1934).
7. M. Gell-Mann, Phys. Rev., **106**, 1296 (1957).
8. R. H. Dalitz, Physical Society (London), Reports on Progress in Physics, **20**, 163 (1957).
9. R. P. Feynman, Phys. Rev., **76**, 749 (1949).
10. R. P. Feynman, Phys. Rev., **74**, 939 (1948).
11. R. P. Feynman, Phys. Rev., **76**, 769 (1949).
12. C. M. Sommerfield, Phys. Rev., **107**, 328 (1957).
13. V. L. Telegdi et al., Phys. Rev., **110**, 1214 (1958).
14. R. P. Feynman and M. Gell-Mann, Phys. Rev., **109**, 193 (1958).

The Fundamental Particles[a]

Particle		Spin	Strangeness	I	I_z	Mass, Mev	Lifetime, sec	Decay mode	Fraction mode
Graviton	g	2				0	Stable		
Photon	γ	1				0	Stable		
Leptons	e^-	1/2				0.510976 ± 0.000007	Stable		
	μ^-	1/2				105.70 ± 0.06	$2.212 \pm 0.001 \times 10^{-6}$	$e^- + \nu + \bar{\nu}$	~ 1
								$e^- + \gamma$	$< 0.7 \times 10^{-6}$ [b]
								$e^- + e^+ + e^-$	1×10^{-5} [c]
								$e^- + \nu + \bar{\nu} + e^+ + e^-$	$1.5 \pm 1.0 \times 10^{-5}$ [c]
	ν	1/2				0	Stable		
Mesons	$\pi^\pm$	0	0	1	±1	139.63 ± 0.06	$(2.56 \pm 0.005) \times 10^{-8}$	$\mu^\pm + \left\{\begin{matrix}\nu \\ \bar{\nu}\end{matrix}\right\}$	~ 1
								$e^\pm \left\{\begin{matrix}\nu \\ \bar{\nu}\end{matrix}\right\}$	$(1.1 \pm 0.3) \times 10^{-4}$ [d]
	π°	0	0	1	0	135.04 ± 0.16	$< 4 \times 10^{-16}$	2γ	1
	K^+	0	1	1/2	1/2	494.0 ± 0.2	$(1.224 \pm 0.013) \times 10^{-8}$	$\pi^+ - \pi^- + \pi^+$	0.062 ± 0.003[e]
								$\pi^+ - \pi^\circ + \pi^\circ$	0.0215 ± 0.003[e]
								$\pi^+ - \pi^\circ$	0.25 ± 0.02[e]
								$\mu^+ + \nu + \pi^\circ$	0.039 ± 0.005[e]
								$e^+ + \nu + \pi^\circ$	0.051 ± 0.008[e]
								$\mu^+ + \nu$	0.58 ± 0.02[e]
	K°	0	1	1/2	−1/2	497.9 ± 0.6	K_1°: $(1.00 \pm 0.038) \times 10^{-10}$	$\pi^\circ + \pi^\circ$	±0.08 } 0.49 ± 0.05[f]
								$\pi^+ + \pi^-$	0.37 }
							K_2°: $6.1^{\;1.6}_{-1.1} \times 10^{-8}$	$\pi^\circ + \pi^\circ + \pi^\circ$	} ~ 0.05[g]
								$\pi^\circ + \pi^+ + \pi^-$	}
								$\mu^\pm + \left\{\begin{matrix}\nu \\ \bar{\nu}\end{matrix}\right\} + \pi^\mp$	~ 0.22[h] } 0.45[g]
								$e^\pm + \left\{\begin{matrix}\nu \\ \bar{\nu}\end{matrix}\right\} + \pi^\mp$	~ 0.22[h] }

(continued)

Particle		Spin	Strangeness	I	I_z	Mass, Mev	Lifetime, sec	Decay mode	Fraction mode
Baryons	γ	1/2	0	1/2	1/2	938.213 ± 0.01	Stable		
	n	1/2	0	1/2	−1/2	939.506 ± 0.01	$(1.04 \pm 0.13) \times 10^3$	$p + e^- + \nu$	1
	Λ	1/2	−1	0	0	1115.36 ± 0.14	$(2.505 \pm 0.086) \times 10^{-10}$	$p + \pi^-$	0.60 ± 0.03[f]
								$n + \pi^\circ$	0.40 ± 0.03[f]
								$p + \epsilon^- + \bar{\nu}$	~ 0.002[f]
								$p + \mu^- + \bar{\nu}$	~ 0.001[f]
	Σ^+	1/2	−1	1	1	1189.40 ± 0.20	$(0.81 \pm 0.06) \times 10^{-10}$	$n + \pi^+$	0.46 ± 0.06[h]
								$p + \pi^\circ$	0.54 ± 0.06[h]
								$n + e^+ + \nu$	~ 0.004[f]
								$n + \mu^+ + \nu$	~ 0.003[f]
								Λ + leptons	< 0.002[f]
	Σ°	1/2	−1	1	0	1191.5 ± 0.5	$< 0.1 \times 10^{-10}$	$\Lambda + \gamma$	1
	Σ^-	1/2	−1	1	−1	1196.0 ± 0.3	$(1.61 \pm 0.1) \times 10^{-10}$	$n + \pi^-$	99.6
								$p + e^- + \bar{\nu}$	~ 0.002[f]
								$p + \mu^- + \bar{\nu}$	~ 0.002[f]
								Λ + leptons	< 0.001[f]
	Ξ°	Fermion	−2	1/2	1/2	1311 ± 8.0	1.5×10^{-10} (1 event)	$\Lambda + \pi^\circ$	1 event
	Ξ^-	Fermion	−2	1/2	−1/2	1318.4 ± 1.2	$1.28 \pm 0.35 \times 10^{-10}$	$\Lambda + \pi^-$	~ 40 events

[a] L. W. Alvarez, "The Interactions of Strange Particles" (1959 Kiev Conference). Also data from the 1960 Rochester Conference on High Energy Physics compiled by W. H. Barkas and A. H. Rosenfeld.

[b] D. Berley, J. Lee, and H. Bardon, Phys. Rev. Letters, **2**, 357-359 (1959).

[c] J. Lee and N. P. Samios, Phys. Rev. Letters, **3**, 55-56 (1959).

[d] H. L. Anderson, T. A. Fujii, R. H. Miller, and L. Tau, Phys. Rev. Letters, **2**, 53-55, 64 (1955).

[e] Steinberger.

[f] D. A. Glazer, "Strange Particle Decays" (1959 Kiev Conference). Data from several sources is reported; not combined.

[g] F. S. Crawford, J. M. Cresti, R. L. Douglass, M. L. Good, G. R. Kalbfleisch, and M. L. Stevenson, Phys. Rev. Letters, **2**, 361-363 (1959).

[h] M. Gell-Mann and A. H. Rosenfeld, Ann. Rev. Nuclear Sci., **7**, 407 (1957).

Milton Keynes UK
Ingram Content Group UK Ltd.
UKHW040054071024
449327UK00019B/554

9 780201 360776